ELEMENTS OF NUMBER THEORY

Including an Introduction to Equations Over Finite Fields

Kenneth Ireland
University of New Brunswick

Michael I. Rosen
Brown University

Bogden & Quigley, Inc.
Publishers

Tarrytown-on-Hudson, New York / Belmont, California

Cover design by Winston G. Potter

Text design by Science Bookcrafters, Inc.

Library of Congress Catalog Card No.: 73-170778

Standard Book No.: 0-8005-0025-3

Printed in the United States of America

1 2 3 4 5 6 7 8 9 10—76 75 74 73 72

PREFACE

There are at present a number of excellent textbooks on the market which introduce the theory of numbers. Some explanation is therefore in order for adding yet another book to this area.

Most of the existing textbooks make very little demand on the algebraic background of the student. It is increasingly common, however, that by the time a student takes a course in number theory, usually in his junior or senior year, he has had at least one course in abstract algebra. It is therefore not unreasonable to build upon this foundation. There are two advantages: the foundational material can be presented in a more systematic, natural way, and the student can be introduced to more advanced theories at an earlier age.

The traditional textbook begins with a discussion of prime numbers, unique factorization, arithmetic functions, congruences, and a presentation of the law of quadratic reciprocity. Thereafter a number of scattered topics are taken up, the selection depending upon the taste of the particular author. We have followed convention, more or less, in the first five chapters. Very little is demanded in the way of background. Nevertheless, it is remarkable how a modicum of group and ring theory introduce unexpected order into the subject. For example, many scattered results turn out to be parts of the answer to a natural question: What is the structure of the group of units in the ring $\mathbb{Z}/n\mathbb{Z}$?

The later chapters draw more heavily on the student's algebraic background, although at no point is more required than the simplest facts about vector spaces, groups, rings, and fields. Able students who have not yet made the acquaintance of these subjects should be able to compensate for this lack by a small amount of supplementary reading.

The content of the later chapters centers about two themes: reciprocity and equations over finite fields. The law of quadratic reciprocity, beautiful in itself, is the first of a number of reciprocity theorems which ultimately culminate in the Artin law of reciprocity, one of the major achievements of the algebraic theory of numbers. We travel on this road beyond quadratic reciprocity by formulating and then proving the law of cubic reciprocity. In preparation for this, many of the techniques used in the general theory are introduced: finite fields, algebraic number fields, splitting of primes, etc. Thus this material provides a stepping stone into the algebraic theory of numbers. It is interesting to note that the new field of algebraic K-theory has provided renewed interest in reciprocity theorems.

The other major theme, equations over finite fields, has a long and honorable history. Gauss, who introduced the notion of congruence (and thereby implicitly finite fields) in his masterpiece, *Disquisitiones Arithmeticae*, contributed many important results. In recent years a great deal of research has been stimulated by a paper of A. Weil, "Number of solutions of equations over finite fields" [80]. One of the major aims of this book is to provide an exposition of a portion of that paper which is sufficiently elementary to be grasped and appreciated by undergraduates. A student who masters this material will have been introduced to an area of mathematics abounding in beautiful theorems and unresolved conjectures.

We have been influenced by many outstanding books during the preparation of this textbook. On the undergraduate level *The Theory of Numbers* by Hardy and Wright and *Introduction to Number Theory* by Niven and Zuckerman were particularly helpful. On the more advanced level *Number Theory* by Borevich and Shafarevich and *Vorlesungen über Zahlentheorie* by Hasse were invaluable.

We would like to express our appreciation to John R. Smart, University of Wisconsin, and John Coates, Harvard University, who read the manuscript and made valuable comments. We also wish to thank Mrs. Edwina Michener, Mr. Steven Galovich, and Mr. Steven Tillman, who read the manuscript carefully and contributed many helpful suggestions, and Mrs. Roberta Weller, Miss Elizabeth Reynolds, and Mrs. Leah King, who did an excellent job of typing it.

KENNETH IRELAND
MICHAEL I. ROSEN

CONTENTS

chapter one/UNIQUE FACTORIZATION

The notion of prime number is fundamental in number theory. This first part of this chapter is devoted to proving that every integer can be written as a product of primes in an essentially unique way.

After that, we shall prove an analogous theorem in the ring of polynomials over a field.

On a more abstract plane, the general idea of unique factorization is treated for principal ideal domains.

Finally, returning from the abstract to the concrete, the general theory is applied to two special rings that will be important later in the book.

I *UNIQUE FACTORIZATION IN Z*

As a first approximation, number theory may be defined as the study of the natural numbers $1, 2, 3, 4, \ldots$. L. Kronecker once remarked (speaking of mathematics generally) that God made the natural numbers and all the rest is the work of man. Although the natural numbers constitute, in some sense, the most elementary mathematical system, the study of their properties has provided generations of mathematicians with problems of unending fascination.

We say that a number a divides a number b if there is a number c such that $b = ac$. If a divides b, we use the notation $a \mid b$. For example, $2 \mid 8$, $3 \mid 15$, but $6 \nmid 21$. If we are given an integer, it is tempting to factor it again and again until further factorization is impossible. For example, $180 = 18 \times 10 = 2 \times 9 \times 2 \times 5 = 2 \times 3 \times 3 \times 2 \times 5$. Numbers that cannot be factored further are called primes. To be more precise, we say that a number p is a prime if its only divisors are 1 and p. Prime numbers are very important because every number can be written as a product of primes. Moreover, primes are of great interest because there are many problems about them that are easy to state but very hard to prove. Indeed many old problems about primes are unsolved to this day.

The first prime numbers are 2, 3, 5, 7, 11, 13, 17, 19, 23, 29, 31, 37, 41, 43, ... One may ask if there are infinitely many prime numbers. The answer is yes. Euclid gave an elegant proof of this fact over 2000 years ago. We shall give his proof and several others in Chapter 2. One can ask other questions of this nature. Let $\pi(x)$ be the number of primes between 1 and x. What can be said about the function $\pi(x)$? Several mathematicians found by experiment that for large x the function $\pi(x)$ was approximately equal to $x/\ln(x)$. This assertion, known as the prime number theorem, was proved toward the end of the nineteenth century by J. Hadamard and independently by Ch.-J. de la Vallé Poussin. More precisely, they proved

$$\lim_{x\to\infty} \frac{\pi(x)}{x/\ln(x)} = 1$$

Even from a small list of primes one can notice that they have a tendency to occur in pairs, for example, 3 and 5, 5 and 7, 11 and 13, 17 and 19. Do there exist infinitely many prime pairs? The answer is unknown.

Another famous unsolved problem is known as the Goldbach conjecture (C. H. Goldbach). Can every even number be written as the sum of two primes? Goldbach came to this conjecture experimentally. Nowadays electronic computers make it possible to experiment with very large numbers. No counterexample to Goldbach's conjecture has ever been found. Great progress toward a proof has been given by I. M. Vinogradov and L. Schnirelmann. In 1937 Vinogradov was able to show that every sufficiently large odd number is the sum of three odd primes.

In this book we shall not go into details about the distribution of prime numbers or into "additive" problems about them (such as the Goldbach conjecture). Rather our concern will be about the way primes enter into the multiplicative structure of numbers. The main theorem along these lines goes back to Euclid. It is the theorem of unique factorization. This theorem is sometimes referred to as the fundamental theorem of arithmetic. It deserves the title. In one way or another almost all the results we shall discuss depend on it. The theorem states that every number can be factored into a product of primes in a unique way. What uniqueness means will be explained below.

As an illustration consider the number 180. We have seen that $180 = 2 \times 2 \times 3 \times 3 \times 5 = 2^2 \times 3^2 \times 5$. Uniqueness in this case means that the only primes dividing 180 are 2, 3, and 5 and that the exponents 2, 2, and 1 are uniquely determined by 180.

$\mathbb{Z}$ will denote the ring of integers, i.e., the set $0, \pm 1, \pm 2, \pm 3, \ldots$, together with the usual definition of sum and product. It will be more convenient to work with $\mathbb{Z}$ rather than restricting ourselves to the positive integers. The notion of divisibility carries over with no difficulty to $\mathbb{Z}$. If p is a positive prime, $-p$ will also be a prime. We shall not consider 1 or -1 as primes even though they fit the definition. This is simply a useful convention. Note that 1 and -1 divide everything and that they are the only integers with this property. They are called the units of $\mathbb{Z}$. Notice also that every integer divides zero. As is usual we shall exclude division by zero.

There are a number of simple properties of division that we shall simply list. The reader may wish to supply the proofs.

1. $a \mid a$.
2. If $a \mid b$ and $b \mid a$, then $a = \pm b$.
3. If $a \mid b$ and $b \mid c$, then $a \mid c$.
4. If $a \mid b$ and $a \mid c$, then $a \mid b + c$.

Let $n \in \mathbb{Z}$ and let p be a prime. Then if n is not zero, there is a non-negative integer a such that $p^a \mid n$ but $p^{a+1} \nmid n$. This is easy to see if both p and n are positive for then the powers of p get larger and larger and eventually exceed n. The other cases are easily reduced to this one. The number a is called the order of n at p and is denoted by $\text{ord}_p\, n$. Roughly speaking $\text{ord}_p\, n$ is the number of times p divides n. If $n = 0$, we set $\text{ord}_p\, 0 = \infty$. Notice that $\text{ord}_p\, n = 0$ if and only if (iff) $p \nmid n$.

Lemma 1

Every nonzero integer can be written as a product of primes.

PROOF

Assume that there is an integer that cannot be written as a product of primes. Let N be the smallest positive integer with this property. Since N cannot itself be prime we must have $N = mn$, where $1 < m, n < N$. However, since m and n are positive and smaller than N they must each be a product of primes. But then so is $N = mn$. This is a contradiction.

The proof can be given in a more positive way by using mathematical induction. It is enough to prove the result for all positive integers. 2 is a prime. Suppose that $2 < N$ and that we have proved the result for all numbers m such that $2 \le m < N$. We wish to show that N is a product of primes. If N is a prime, there is nothing to do. If N is not a prime, then $N = mn$, where $2 \le m, n < N$. By induction both m and n are products of primes and thus so is N.

By collecting terms we can write $n = p_1^{a_1} p_2^{a_2} \cdots p_m^{a_m}$, where the p are primes and the a_i are nonnegative integers. We shall use the following notation:

$$n = (-1)^{\varepsilon(n)} \prod_p p^{a(p)}$$

where $\varepsilon(n) = 0$ or 1 depending on whether n is positive or negative and where the product is over all positive primes. The exponents $a(p)$ are nonnegative integers and, of course, $a(p) = 0$ for all but finitely many primes. For example, if $n = 180$, we have $\varepsilon(n) = 0$, $a(2) = 2$, $a(3) = 2$, and $a(5) = 1$, and all other $a(p) = 0$.

We can now state the main theorem.

Theorem 1

For every nonzero integer n there is a prime factorization

$$n = (-1)^{\varepsilon(n)} \prod_p p^{a(p)}$$

with the exponents uniquely determined by n. In fact, we have $a(p) = \operatorname{ord}_p n$.

The proof of this theorem is not as easy as it may seem. We shall postpone the proof until we have established a few preliminary results.

Lemma 2

If $a, b \in \mathbb{Z}$ *and* $a > b$, *there exist* $q, r \in \mathbb{Z}$ *such that* $a = qb + r$ *with* $0 \leq r < b$.

PROOF

Consider the set of all integers of the form $a - xb$ with $x \in \mathbb{Z}$. This set includes positive elements. Let $r = a - qb$ be the least positive element in this set. We claim that $0 \leq r < b$. If not, $r = a - qb \geq b$ and so $0 \leq a - (q + 1)b < r$, which contradicts the minimality of r.

Definition

If $a_1, a_2, \ldots, a_n \in \mathbb{Z}$, we define $(a_1, a_2, \ldots, a_n)$ to be the *set of all integers* of the form $a_1x_1 + a_2x_2 + \cdots + a_nx_n$ with $x_1, x_2, \ldots, x_n \in \mathbb{Z}$.

Let $A = (a_1, a_2, \ldots, a_n)$. Notice that the sum and difference of two elements in A are again in A. Also, if $a \in A$ and $r \in \mathbb{Z}$, then $ra \in A$. In ring-theoretic language, A is an *ideal* in the ring Z.

Lemma 3

If $a, b \in \mathbb{Z}$, *then there is a* $d \in \mathbb{Z}$ *such that* $(a, b) = (d)$.

PROOF

We may assume that not both a and b are zero and so that there are positive elements in (a, b). Let d be the smallest positive element in (a, b). Clearly $(d) \subseteq (a, b)$. We shall show that the reverse inclusion also holds.

Suppose that $c \in (a, b)$. By Lemma 2 there exist integers q and r such that $c = qd + r$ with $0 \leq r < d$. Since both c and d are in (a, b) it follows that $r = c - qd$ is also in (a, b). Since $0 \leq r < d$ we must have $r = 0$. Thus $c = qd \in (d)$.

Definition

Let $a, b \in \mathbb{Z}$. An integer d is called a *greatest common divisor* of a and b if d is a divisor of both a and b and if every other common divisor of a and b divides d.

Notice that if c is another greatest common divisor of a and b, then we must have $c \mid d$ and $d \mid c$ and so $c = \pm d$. Thus the greatest common divisor of two numbers, if it exists, is determined up to sign.

As an example, one may check that 14 is a greatest common divisor of 42 and 196. The following lemma will establish the existence of the greatest common divisor, but it will not give a method for computing it. In the Exercises we shall outline an efficient method of computation known as the Euclidean algorithm.

Lemma 4

Let $a, b \in \mathbb{Z}$. By Lemma 3 we have $(a, b) = (d)$. d is a greatest common divisor of a and b.

PROOF

Since $a \in (d)$ and $b \in (d)$ we see that d is a common divisor of a and b. Suppose that c is a common divisor. Then c divides every number of the form $ax + by$. In particular $c \mid d$.

Definition

We say that two integers a and b are *relatively prime* if the only common divisors are ± 1, the units.

It is fairly standard to use the notation (a, b) for the greatest common divisor of a and b. The way we have defined things, (a, b) is a set. However, since $(a, b) = (d)$ and d is a greatest common divisor (if we require d to be positive, we may use the article *the*) it will not be too confusing to use the symbol (a, b) for both meanings. With this convention we can say that a and b are relatively prime if $(a, b) = 1$.

Proposition 1.1.1

Suppose that $a \mid bc$ and that $(a, b) = 1$. Then $a \mid c$.

PROOF

Since $(a, b) = 1$ there exist integers r and s such that $ra + sb = 1$. Therefore, $rac + sbc = c$. Since a divides the left-hand side of this equation we have $a \mid c$.

This proposition is false if $(a, b) \neq 1$. For example, $6 \mid 24$ but $6 \nmid 3$ and $6 \nmid 8$.

Corollary 1

If p is a prime and $p \mid bc$, then either $p \mid b$ or $p \mid c$.

PROOF

The only divisors of p are ± 1 and $\pm p$. Thus $(p, b) = 1$ or p; i.e., either $p \mid b$ or p and b are relatively prime. If $p \mid b$, we are done. If not, $(p, b) = 1$ and so, by the proposition, $p \mid c$.

We can state the corollary in a slightly different form that is often useful: If p is a prime and $p \nmid b$ and $p \nmid c$, then $p \nmid bc$.

Corollary 2

Suppose that p is a prime and that $a, b \in \mathbb{Z}$. Then $\operatorname{ord}_p ab = \operatorname{ord}_p a + \operatorname{ord}_p b$.

PROOF

Let $\alpha = \operatorname{ord}_p a$ and $\beta = \operatorname{ord}_p b$. Then $a = p^\alpha c$ and $b = p^\beta d$, where $p \nmid b$ and $p \nmid c$. Then $ab = p^{\alpha+\beta} cd$ and by Corollary 1 $p \nmid cd$. Thus $\operatorname{ord}_p ab = \alpha + \beta = \operatorname{ord}_p a + \operatorname{ord}_p b$.

We are now in a position to prove the main theorem.

Apply the function ord_q to both sides of the equation

$$n = (-1)^{\varepsilon(n)} \prod_p p^{a(p)}$$

and use the property of ord_q given by Corollary 2. The result is

$$\operatorname{ord}_q n = \varepsilon(n) \operatorname{ord}_q(-1) + \sum_q a(p) \operatorname{ord}_q(p)$$

Now, from the definition of ord_q we have $\operatorname{ord}_q(-1) = 0$ and $\operatorname{ord}_q(p) = 0$ if $p \neq q$ and 1 if $p = q$. Thus the right-hand side collapses to the single term $a(q)$, i.e., $\operatorname{ord}_q n = a(q)$, which is what we wanted to prove.

It is to be emphasized that the key step in the proof is Corollary 1: namely, if $p \mid ab$, then $p \mid a$ or $p \mid b$. Whatever difficulty there is in the proof is centered about this fact.

2 UNIQUE FACTORIZATION IN $k[x]$

The theorem of unique factorization can be formulated and proved in more general contexts than that of Section 1. In this section we shall consider the ring $k[x]$ of polynomials with coefficients in a field k. In Section 3 we shall consider principal ideal domains. It will turn out that the analysis of these situations will prove useful in the study of the integers.

If $f, g \in k[x]$, we say that f divides g if there is an $h \in k[x]$ such that $g = fh$.

If $\deg f$ denotes the degree of f, we have $\deg fg = \deg f + \deg g$. Also, remember that $\deg f = 0$ iff f is a nonzero constant. It follows that $f \mid g$ and $g \mid f$ iff $f = cg$, where c is a nonzero constant. It also follows that the only polynomials that divide all the others are the nonzero constants. These are the units of $k[x]$. A nonconstant polynomial p is said to be irreducible if $q \mid p$ implies that q is either a constant or a constant times p. Irreducible polynomials are the analog of prime numbers.

Lemma 1

Every nonconstant polynomial is the product of irreducible polynomials.

PROOF

The proof is by induction on the degree. It is easy to see that polynomials of degree 1 are irreducible. Assume that we have proved the result for all polynomials of degree less than n and that $\deg f = n$. If f is irreducible, we are done. Otherwise $f = gh$, where $1 \leq \deg g$, $\deg h < n$. By the induction assumption both g and h are products of irreducible polynomials. Thus so is $f = gh$.

It is convenient to define *monic polynomial*. A polynomial f is called monic if its leading coefficient is 1. For example, $x^2 + x - 3$ and $x^3 - x^2 + 3x + 17$ are monic but $2x^3 - 5$ and $3x^4 + 2x^2 - 1$ are not. Every polynomial (except zero) is a constant times a monic polynomial.

Let p be a monic irreducible polynomial. We define $\operatorname{ord}_p f$ to be the integer a defined by the property that $p^a \mid f$ but that $p^{a+1} \nmid f$. Such an integer must exist since the degree of the powers of p gets larger and larger. Notice that $\operatorname{ord}_p f = 0$ iff $p \nmid f$.

Theorem 2

Let $f \in k[x]$. Then we can write

$$f = c \prod_p p^{a(p)}$$

where the product is over all monic irreducible polynomials and c is a constant. The constant c and the exponents a(p) are uniquely determined by f; in fact, $a(p) = \operatorname{ord}_p f$.

The existence of such a product follows immediately from Lemma 1. As before, the uniqueness is more difficult and the proof will be postponed until we develop a few tools.

Lemma 2

Let $f, g \in k[x]$. *If* $g \neq 0$, *there exist polynomials* $h, r \in k[x]$ *such that* $f = hg + r$, *where either* $r = 0$ *or* $r \neq 0$ *and* $\deg r < \deg g$.

PROOF

If $g \mid f$, simply set $h = f/g$ and $r = 0$. If $g \nmid f$, let $r = f - hg$ be the polynomial of least degree among all polynomials of the form $f - lg$ with $l \in k[x]$. We claim that $\deg r < \deg g$. If not, let the leading term of r be ax^d and that of g be bx^m. Then $r - ab^{-1}x^{d-m}g = f - (h + ab^{-1}x^{d-m})g$ has smaller degree than r and is of the given form. This is a contradiction.

Definition

If $f_1, f_2, \ldots, f_n \in k[x]$, then $(f_1, f_2, \ldots, f_n)$ is the *set of all polynomials* of the form $f_1h_1 + f_2h_2 + \cdots + f_nh_n$, where $h_1, h_2, \ldots, h_n \in k[x]$.

In ring-theoretic language $(f_1, f_2, \ldots, f_n)$ is the ideal generated by $f_1, f_2, \ldots, f_n$.

Lemma 3

Given $f, g \in k[x]$ *there is a* $d \in k[x]$ *such that* $(f, g) = (d)$.

PROOF

In the set (f, g) let d be an element of least degree. We have $(d) \subseteq (f, g)$ and we want to prove the reverse inclusion. Let $c \in (f, g)$. If $d \nmid c$, then there exist polynomials h and r such that $c = hd + r$ with $\deg r < \deg d$. Since c and d are in (f, g) we have $r = c - hd \subseteq (f, g)$. Since r has smaller degree than d this is a contradiction. Therefore, $d \mid c$ and $c \in (d)$.

Definition

Let $f, g \in k[x]$. Then $d \in k[x]$ is said to be a *greatest common divisor* of f and g if d divides f and g and every common divisor of f and g divides d.

Notice that the greatest common divisor of two polynomials is determined up to multiplication by a constant. If we require it to be monic, it is uniquely determined and we may speak of *the* greatest common divisor.

Lemma 4

Let $f, g \in k[x]$. By Lemma 3 there is a $d \in k[x]$ such that $(f, g) = (d)$. d is a greatest common divisor of f and g.

PROOF

Since $f \in (d)$ and $g \in (d)$ we have $d \mid f$ and $d \mid g$. Suppose that $h \mid f$ and that $h \mid g$. Then h divides every polynomial of the form $fl + gm$ with $l, m \in k[x]$. In particular $h \mid d$, and we are done.

Definition

Two polynomials f and g are said to be *relatively prime* if the only common divisors of f and g are constants. In other words, $(f, g) = (1)$.

Proposition 1.2.1

If f and g are relatively prime and $f \mid gh$, then $f \mid h$.

PROOF

If f and g are relatively prime, we have $(f, g) = (1)$ so there are polynomials l and m such that $lf + mg = 1$. Thus $lfh + mgh = h$. Since f divides the left-hand side of this equation f must divide h.

Corollary 1

If p is an irreducible polynomial and $p \mid fg$, then $p \mid f$ or $p \mid g$.

PROOF

Since p is irreducible $(p, f) = (p)$ or (1). In the first case $p \mid f$ and we are done. In the second case p and f are relatively prime and the result follows from the proposition.

Corollary 2

If p is a monic irreducible polynomial and $f, g \in k[x]$, we have $\operatorname{ord}_p fg = \operatorname{ord}_p f + \operatorname{ord}_p g$.

PROOF

The proof is almost word for word the same as the proof to Corollary 2 to Proposition 1.1.1.

The proof of Theorem 2 is now easy. Apply the function ord_q to both sides of the relation

$$f = c \prod_p p^{a(p)}$$

We find that

$$\text{ord}_q f = \text{ord}_q c + \sum_q a(p)\, \text{ord}_q p$$

Now, since c is a constant $q \nmid c$ and $\text{ord}_q c = 0$. Moreover, $\text{ord}_q p = 0$ if $q \neq p$ and 1 if $q = p$. Thus the above relation yields $\text{ord}_q f = a(q)$. This shows that the exponents are uniquely determined. It is clear that if the exponents are uniquely determined by f, then so is c. This completes the proof.

3 *UNIQUE FACTORIZATION IN A PRINCIPAL IDEAL DOMAIN*

The reader will not have failed to notice the great similarity in the methods of proof in Sections 1 and 2. In this section we shall prove an abstract theorem that includes the previous results as special cases.

Throughout this section R will denote an integral domain.

Definition 1

R is said to be a *Euclidean domain* if there is a function λ from the nonzero elements of R to the set $\{0, 1, 2, 3, \ldots\}$ such that if $a, b \in R, b \neq 0$, there exists $c, d \in R$ with the property $a = cb + d$ and either $d = 0$ or $\lambda(d) < \lambda(b)$.

The rings $\mathbb{Z}$ and $k[x]$ are both Euclidean domains. In $\mathbb{Z}$ we can take ordinary absolute value as the function λ; in the ring $k[x]$ the function that assigns to every polynomial its degree will serve the purpose.

Proposition 1.3.1

If R is a Euclidean domain and $I \subseteq R$ is an ideal, then there is an element $a \in R$ such that $I = Ra = \{ra \mid r \in R\}$.

PROOF

Consider the set of nonnegative integers $\{\lambda(b) \mid b \in I\ b \neq 0\}$. Since every set of nonnegative integers has a least element there is an $a \in I$, $a \neq 0$, such that $\lambda(a) \leq \lambda(b)$ for all $b \in I$, $b \neq 0$. We claim that $I = Ra$. Clearly, $Ra \subseteq I$. Suppose that $b \in I$; then we know that there are elements $c, d \in R$ such that $b = ca + d$, where either $d = 0$ or $\lambda(d) < \lambda(a)$. Since $d = b - ca \in I$ we cannot have $\lambda(d) < \lambda(a)$. Thus $d = 0$ and $b = ca \in Ra$. Therefore, $I \subseteq Ra$ and we are done.

For elements $a_1, \ldots, a_n \in R$, define $(a_1, a_2, \ldots, a_n) = Ra_1 + Ra_2 + \cdots + Ra_n = \{\sum_{i=1}^{n} r_i a_i | r_i \in R\}$. $(a_1, a_2, \ldots, a_n)$ is an ideal. If an ideal I is equal to $(a_1, \ldots, a_n)$ for some elements $a_i \in I$, we say that I is finitely generated. If $I = (a)$ for some $a \in I$, we say that I is a principal ideal.

Definition 2
R is said to be a *principal ideal domain* (PID) if every ideal of R is principal.

Proposition 1.3.1 asserts that every Euclidean domain is a PID. The converse of this statement is false, although it is somewhat hard to provide examples.

The remaining discussion in this section is about PID's. The notion of Euclidean domain is useful because in practice one can show that many rings are PID's by first establishing that they are Euclidean domains. We shall give two further examples in Section 4.

We introduce some more terminology. If $a, b \in R$, $b \neq 0$, we say that b divides a if $a = bc$ for some $c \in R$. Notation: $b \mid a$. An element $u \in R$ is called a unit if u divides 1. Two elements $a, b \in R$ are said to be associates if $a = bu$ for some unit u. An element $p \in R$ is said to be irreducible if $a \mid p$ implies that a is either a unit or an associate of p. A nonunit $p \in R$ is said to be prime if $p \neq 0$ and $p \mid ab$ implies that $p \mid a$ or $p \mid b$.

The distinction between irreducible element and prime element is new. In general these notions do not coincide. As we have seen they do coincide in Z and $k[x]$, and we shall prove shortly that they coincide in a PID.

Some of the notions we are discussing can be translated into the language of ideals. Thus $a \mid b$ iff $(b) \subseteq (a)$. $u \in R$ is a unit iff $(u) = R$. a and b are associate iff $(a) = (b)$. p is prime iff $ab \in (p)$ implies that either $a \in (p)$ or $b \in (p)$. All these assertions are easy exercises. The notion of irreducible element can be formulated in terms of ideals, but we will not need it.

Definition
$d \in R$ is said to be a *greatest common divisor* (gcd) of two elements $a, b \in R$ if

(a) $d \mid a$ and $d \mid b$.

(b) $d' \mid a$ and $d' \mid b$ implies that $d' \mid d$.

It is easy to see that if both d and d' are gcd's of a and b, then d is associate to d'.

The gcd of two elements need not exist in a general ring. However,

Proposition 1.3.2
Let R be a PID *and $a, b \in R$. Then a and b have a greatest common divisor d and $(a, b) = (d)$.*

PROOF
Form the ideal (a, b). Since R is a PID there is an element d such that $(a, b) = (d)$. Since $(a) \subseteq (d)$ and $(b) \subseteq (d)$ we have $d \mid a$ and $d \mid b$. If $d' \mid a$ and $d' \mid b$, then $(a) \subseteq (d')$ and $(b) \subseteq (d')$. Thus $(d) = (a, b) \subseteq (d')$ and $d' \mid d$. We have proved that d is a gcd of a and b and that $(a, b) = (d)$.

Two elements a and b are said to be relatively prime if the only common divisors are units.

Corollary 1
If R is a PID *and $a, b \in R$ are relatively prime, then $(a, b) = R$.*

Corollary 2
If R is a PID *and $p \in R$ is irreducible, then p is prime.*

PROOF
Suppose that $p \mid ab$ and that $p \nmid a$. Since $p \nmid a$ it follows that the only common divisors are units. By Corollary 1 $(a, p) = R$. Thus $(ab, pb) = (b)$. Since $ab \in (p)$ and $pb \in (p)$ we have $(b) \subseteq (p)$. Thus $p \mid b$.
It is easy to see that a prime is irreducible.

From now on R will be a PID and we shall use the words *prime* and *irreducible* interchangeably.
We want to show that every nonzero element of R is a product of irreducible elements. The proof is in two steps. First one shows that if $a \in R$, $a \neq 0$, there is an irreducible dividing a. Then we show that a is a product of irreducibles.

Lemma 1
Let $(a_1) \subseteq (a_2) \subseteq (a_3) \subseteq \cdots$ be an ascending chain of ideals. Then there is an integer k such that $(a_k) = (a_{k+l})$ for $l = 0, 1, 2, \ldots$. In other words, the chain breaks off in finitely many steps.

PROOF
Let $I = \bigcup_{i=1}^{\infty} (a_i)$. It is easy to see that I is an ideal. Thus $I = (a)$ for some $a \in R$. But $a \in \bigcup_{i=1}^{\infty} (a_i)$ implies that $a \in (a_k)$ for some k, which shows that $I = (a) \subseteq (a_k)$. It follows that $I = (a_k) = (a_{k+1}) = \cdots$.

Proposition 1.3.3
Every nonzero nonunit of R is a product of irreducibles.

PROOF

Let $a \in R$, $a \neq 0$, a not a unit. We wish to show, to begin with, that a is divisible by an irreducible element. If a is irreducible, we are done. Otherwise $a = a_1 b_1$, where a_1 and b_1 are nonunits. If a_1 is irreducible, we are done. Otherwise $a_1 = a_2 b_2$, where a_2 and b_2 are nonunits. If a_2 is irreducible, we are done. Otherwise continue as before. Notice that $(a) \subset (a_1) \subset (a_2) \subset \cdots$. By Lemma 1 this chain cannot go on indefinitely. Thus for some k, a_k is irreducible.

We now show that a is a product of irreducibles. If a is irreducible, we are done. Otherwise let p_1 be an irreducible such that $p_1 \mid a$. Then $a = p_1 c_1$. If c_1 is a unit, we are done. Otherwise let p_2 be an irreducible such that $p_2 \mid c_1$. Then $a = p_1 p_2 c_2$. If c_2 is a unit, we are done. Otherwise continue as before. Notice that $(a) \subset (c_1) \subset (c_2) \subset \cdots$. This chain cannot go on indefinitely by Lemma 1. Thus for some k, $a = p_1 p_2 \cdots p_k c_k$, where c_k is a unit. Since $p_k c_k$ is irreducible, we are done.

We now want to define an ord function as we have done in Sections 1 and 2.

Lemma 2

Let p be a prime and $a \neq 0$. Then there is an integer n such that $p^n \mid a$ but $p^{n+1} \nmid a$.

PROOF

If the lemma were false, then for each integer $m > 0$ there would be an element b_m such that $a = p^m b_m$. Then $pb_{m+1} = b_m$ so that $(b_1) \subset (b_2) \subset (b_3) \subset \cdots$ would be an infinite ascending chain of ideals that does not break off. This contradicts Lemma 1.

The integer n, which is defined in Lemma 2, is uniquely determined by p and a. We set $n = \operatorname{ord}_p a$.

Lemma 3

If $a, b \in R$ with $a, b \neq 0$, then $\operatorname{ord}_p ab = \operatorname{ord}_p a + \operatorname{ord}_p b$.

PROOF

Let $\alpha = \operatorname{ord}_p a$ and $\beta = \operatorname{ord}_p b$. Then $a = p^\alpha c$ and $b = p^\beta d$ with $p \nmid c$ and $p \nmid d$. Thus $ab = p^{\alpha+\beta} cd$. Since p is prime $p \nmid cd$. Consequently, $\operatorname{ord}_p ab = \alpha + \beta = \operatorname{ord}_p a + \operatorname{ord}_p b$.

We are now in a position to formulate and prove the main theorem of this section.

Let S be a set of primes in R with the following two properties:
(a) Every prime in R is associate to a prime in S.
(b) No two primes in S are associate.

To obtain such a set choose one prime out of each class of associate primes. There is clearly a great deal of arbitrariness in this choice. In Z and $k[x]$ there were natural ways to make the choice. In Z we chose S to be the set of positive primes. In $k[x]$ we chose S to be the set of monic irreducible polynomials. In general there is no neat way to make the choice and this occasionally leads to complications (see Chapter 9).

Theorem 3
Let R be a PID *and S a set of primes with the properties given above. Then if $a \in R$, $a \neq 0$, we can write*

$$a = u \prod_p p^{e(p)} \tag{1}$$

where u is a unit and the product is over all $p \in S$. The unit u and the exponents $e(p)$ are uniquely determined by a. In fact, $e(p) = \operatorname{ord}_p a$.

PROOF
The existence of such a decomposition follows immediately from Proposition 1.3.3.

To prove the uniqueness, let q be a prime in S and apply ord_q to both sides of Equation (1). Using Lemma 3 we get

$$\operatorname{ord}_q a = \operatorname{ord}_q u + \sum_q e(p) \operatorname{ord}_q p$$

Now, from the definition of ord_q we see that $\operatorname{ord}_q u = 0$ and that $\operatorname{ord}_q p = 0$ if $q \neq p$ and 1 if $q = p$. Thus $\operatorname{ord}_q a = e(q)$. Since the exponents $e(q)$ are uniquely determined so is the unit u. This completes the proof.

4 *THE RINGS* $\mathbb{Z}[i]$ *AND* $\mathbb{Z}[\omega]$

As an application of the results in Section 3 we shall consider two examples that will be useful to us in later chapters.

Let $i = \sqrt{-1}$ and consider the set of complex numbers $\mathbb{Z}[i]$ defined by $\{a + bi \mid a, b \in \mathbb{Z}\}$. This set is clearly closed under addition and subtraction. Moreover, if $a + bi, c + di \in \mathbb{Z}$, then $(a + bi)(c + di) =$

$ac + adi + bci + bdi^2 = (ac - bd) + (ad + bc)i \in \mathbb{Z}[i]$. Thus $\mathbb{Z}[i]$ is closed under multiplication and is a ring. Since $\mathbb{Z}[i]$ is contained in the complex numbers it is an integral domain.

Proposition 1.4.1
$\mathbb{Z}[i]$ is a Euclidean domain.

PROOF
For $a + bi \in \mathbb{Z}[i]$ define $\lambda(a + bi) = a^2 + b^2$.

Let $\alpha = a + bi$ and $\gamma = c + di$ and suppose that $\gamma \neq 0$. $\alpha/\gamma = r + si$, where r and s are real numbers (they are, in fact, rational). Choose integers $m, n \in \mathbb{Z}$ such that $|r - m| \leq \frac{1}{2}$ and $|s - n| \leq \frac{1}{2}$. Set $\delta = m + ni$. Then $\delta \in \mathbb{Z}[i]$ and $\lambda((\alpha/\gamma) - \delta) = (r - m)^2 + (s - n)^2 \leq \frac{1}{4} + \frac{1}{4} = \frac{1}{2}$. Set $\rho = \alpha - \gamma\delta$. Then $\rho \in \mathbb{Z}[i]$ and either $\rho = 0$ or $\lambda(\rho) = \lambda(\gamma((\alpha/\gamma) - \delta)) = \lambda(\gamma)\lambda((\alpha/\gamma) - \delta) \leq \frac{1}{2}\lambda(\gamma) < \lambda(\gamma)$.

It follows that λ makes $\mathbb{Z}[i]$ into a Euclidean domain.

The ring $\mathbb{Z}[i]$ is called the ring of Gaussian integers after C. F. Gauss, who first studied its arithmetic properties in detail.

The numbers $\pm 1, \pm i$ are the roots of $x^4 = 1$ over the complex numbers. Consider the equation $x^3 = 1$. Since $x^3 - 1 = (x - 1) \times (x^2 + x + 1)$ the roots of this equation are $1, (-1 \pm \sqrt{-3})/2$. Let $\omega = (-1 + \sqrt{-3})/2$. Then it is easy to check that $\omega^2 = (-1 - \sqrt{-3})/2$ and that $1 + \omega + \omega^2 = 0$.

Consider the set $\mathbb{Z}[\omega] = \{a + b\omega \mid a, b \in Z\}$. $Z[\omega]$ is closed under addition and subtraction. Moreover, $(a + b\omega)(c + d\omega) = ac + (ad + bc)\omega + bd\omega^2 = (ac - bd) + (ad + bc - bd)\omega$. Thus $\mathbb{Z}[\omega]$ is a ring. Again, since $\mathbb{Z}[\omega]$ is a subset of the complex numbers it is an integral domain.

We remark that $\mathbb{Z}[\omega]$ is closed under complex conjugation. In fact, since $\overline{\sqrt{-3}} = \overline{\sqrt{3}i} = -\sqrt{3}i = -\sqrt{-3}$ we see that $\bar{\omega} = \omega^2$. Thus if $\alpha = a + b\omega \in \mathbb{Z}[\omega]$, then $\bar{\alpha} = a + b\bar{\omega} = a + b\omega^2 = (a - b) + b\omega \in \mathbb{Z}[\omega]$.

Proposition 1.4.2
$\mathbb{Z}[\omega]$ is a Euclidean domain.

PROOF
For $\alpha = a + b\omega \in \mathbb{Z}[\omega]$ define $\lambda(\alpha) = a^2 - ab + b^2$. A simple calculation shows that $\lambda(\alpha) = \alpha\bar{\alpha}$.

Now, let $\alpha, \beta \in \mathbb{Z}[\omega]$ and suppose that $\beta \neq 0$. Then $\alpha/\beta = \alpha\bar{\beta}/\beta\bar{\beta} = r + s\omega$, where r and s are rational numbers. We have used the fact that $\beta\bar{\beta} = \lambda(\beta)$ is a positive integer and that $\alpha\bar{\beta} \in \mathbb{Z}[\omega]$ since α and $\bar{\beta} \in \mathbb{Z}[\omega]$.

Find integers m and n such that $|r - m| \leq \frac{1}{2}$ and $|s - n| \leq \frac{1}{2}$. Then put $\gamma = m + n\omega$. $\lambda((\alpha/\beta) - \gamma) = (r - m)^2 - (r - m)(s - n) + (s - n)^2 \leq \frac{1}{4} + \frac{1}{4} + \frac{1}{4} < 1$.

Let $\rho = \alpha - \gamma\beta$. Then either $\rho = 0$ or $\lambda(\rho) = \lambda(\beta((\alpha/\beta) - \gamma)) = \lambda(\beta)\lambda((\alpha/\beta) - \gamma) < \lambda(\beta)$.

From the analysis of Section 3 we know that the theorem of unique factorization is true in both $\mathbb{Z}[i]$ and $\mathbb{Z}[\omega]$. To go further with the analysis of these rings we would have to investigate the units and the prime elements. There are some results of this nature in the exercises.

Notes

Rings for which the theorem of unique factorization into primes holds are called unique factorization domains (UFD). The fact that Z is a UFD is already implicit in Euclid, but the first explicit and clear statement of the result seems to be in C. F. Gauss's masterpiece *Disquisitiones Arithmeticae* (available in English translation by A. A. Clark, Yale University Press, New Haven, Conn., 1966). Zermelo gave a clever proof by contradiction, which is reproduced in the excellent book of G. H. Hardy and Wright [40].

We have shown that every PID is a UFD. The converse is not true. For example, the ring of polynomials over a field in more than one variable is a UFD but not a PID. P. Samuel has an excellent expository article on UFD's in [67]. A more elementary introduction may be found in the book of H. Rademacher and O. Toeplitz [65].

The reader may find it profitable to read the introductory material in several books on number theory. Chapter 3 of A. Frankel [32] and the introduction to H. Stark [73] are particularly good. There is also an early lecture by Hardy [39] that is highly recommended.

The ring $Z[i]$ was introduced by Gauss in his second memoir on biquadratic reciprocity [34]. G. Eisenstein considered the ring $Z[\omega]$ in connection with his work on cubic reciprocity. He mentions that to investigate the properties of this ring one need only consult Gauss's work on $Z[i]$ and modify the proofs [28]. A thorough treatment of these two rings is given in Chapter 12 of Hardy and Wright [40]. In Chapter 14 they treat a generalization, namely, rings of integers in quadratic number fields. Stark's Chapter 8 deals with the same subject [73]. In 1966 Stark resolved a long-outstanding problem in the theory of numbers by showing that the ring of integers (see Chapter 6 of this book) in the field $Q(\sqrt{d})$, with d negative, is a UFD when $d = -1, -2, -3, -7, -11, -19, -43, -67$, and -163 and for no other values of d.

The student who is familiar with a little algebra will notice that a "generic" non-UFD is given by the ring $k[x, y, z, w]$, with $xy = zw$, where k is a field. Another example of a non-UFD is $\mathbb{C}[x, y, z]$, with $x^2 + y^2 + z^2 = 1$, where $\mathbb{C}$ is the field of complex numbers. To see this notice that $(x + iy)(x - iy) = (1 - z)(1 + z)$.

Exercises

1. Let a and b be nonzero integers. We can find nonzero integers q and r such that $a = qb + r$, where $0 \leq r < b$. Prove that $(a, b) = (b, r)$.
2. (continuation) If $r \neq 0$, we can find q_1 and r_1 such that $b = q_1 r + r_1$ with $0 \leq r_1 < r$. Show that $(a, b) = (r, r_1)$. This process can be repeated. Show that it must end in finitely many steps. Show that the last nonzero remainder must equal (a, b). The process looks like

$$\begin{aligned} a &= qb + r & 0 &\leq r < b \\ b &= q_1 r + r_1 & 0 &\leq r_1 < r \\ r &= q_2 r_1 + r_2 & 0 &\leq r_2 < r_1 \\ &\vdots \\ r_{k-1} &= q_{k+1} r_k + r_{k+1} & 0 &\leq r_{k+1} < r_k \\ r_k &= q_{k+2} r_{k+1} \end{aligned}$$

   Then $r_{k+1} = (a, b)$. This process of finding (a, b) is known as the Euclidean algorithm.
3. Calculate (187, 221), (6188, 4709), and (314, 159).
4. Let $d = (a, b)$. Show how one can use the Euclidean algorithm to find numbers m and n such that $am + bn = d$. (*Hint:* In Exercise 2 we have that $d = r_{k+1}$. Express r_{k+1} in terms of r_k and r_{k-1}, then in terms of r_{k-1} and r_{k-2}, etc.)
5. Find m and n for the pairs a and b given in Exercise 3.
6. Let $a, b, c \in \mathbb{Z}$. Show that the equation $ax + by = c$ has solutions in integers iff $(a, b) \mid c$.
7. Let $d = (a, b)$ and $a = da'$ and $b = db'$. Show that $(a', b') = 1$.
8. Let x_0 and y_0 be a solution to $ax + by = c$. Show that all solutions have the form $x = x_0 + t(b/d)$, $y = y_0 - t(a/d)$, where $d = (a, b)$ and $t \in \mathbb{Z}$.
9. Suppose that $u, v \in \mathbb{Z}$ and that $(u, v) = 1$. If $u \mid n$ and $v \mid n$, show that $uv \mid n$. Show that this is false if $(u, v) \neq 1$.
10. Suppose that $(u, v) = 1$. Show that $(u + v, u - v)$ is either 1 or 2.
11. Show that $(a, a + k) \mid k$.
12. Suppose that we take several copies of a regular polygon and try to fit them evenly about a common vertex. Prove that the only possibilities are six equilateral triangles, four squares, and three hexagons.
13. Let $n_1, n_2, \ldots, n_s \in \mathbb{Z}$. Define the greatest common divisor d of $n_1, n_2, \ldots, n_s$ and prove that there exist integers $m_1, m_2, \ldots, m_s$ such that $n_1 m_1 + n_2 m_2 + \cdots + n_s m_s = d$.
14. Discuss the solvability of $a_1 x_1 + a_2 x_2 + \cdots + a_s x_s = c$ in integers. (*Hint:* Use Exercise 13 to extend the reasoning behind Exercise 6.)
15. Prove that $a \in \mathbb{Z}$ is the square of another integer iff $\text{ord}_p\, a$ is even for all primes p. Give a generalization.

16 If $(u, v) = 1$ and $uv = a^2$, show that both u and v are squares.

17 Prove that the square root of 2 is irrational, i.e., that there is no rational number $r = a/b$ such that $r^2 = 2$.

18 Prove that $\sqrt[n]{m}$ is irrational if m is not the nth power of an integer.

19 Define the least common multiple of two integers a and b to be an integer m such that $a \mid m$, $b \mid m$, and m divides every common multiple of a and b. Show that such an m exists. It is determined up to sign. We shall denote it by $[a, b]$.

20 Prove the following:
(a) $\text{ord}_p[a, b] = \max(\text{ord}_p a, \text{ord}_p b)$.
(b) $(a, b)[a, b] = ab$.
(c) $(a + b, [a, b]) = (a, b)$.

21 Prove that $\text{ord}_p(a + b) \geq \min(\text{ord}_p a, \text{ord}_p b)$ with equality holding if $\text{ord}_p a \neq \text{ord}_p b$.

22 Almost all the previous exercises remain valid if instead of the ring $\mathbb{Z}$ we consider the ring $k[x]$. Indeed, in most we can consider any Euclidean domain. Convince yourself of this fact. For simplicity we shall continue to work in $\mathbb{Z}$.

23 Suppose that $a^2 + b^2 = c^2$ with $a, b, c \in \mathbb{Z}$. For example, $3^2 + 4^2 = 5^2$ and $5^2 + 12^2 = 13^2$. Assume that $(a, b) = (b, c) = (c, a) = 1$. Prove that there exist integers u and v such that $c - b = 2u^2$ and $c + b = 2v^2$ and $(u, v) = 1$ (there is no loss in generality in assuming that b and c are odd and that a is even). Consequently $a = 2uv$, $b = v^2 - u^2$, and $c = v^2 + u^2$. Conversely show that if u and v are given, then the three numbers a, b, and c given by these formulas satisfy $a^2 + b^2 = c^2$.

24 Prove the identities
(a) $x^n - y^n = (x - y)(x^{n-1} + x^{n-2}y + \cdots + y^{n-1})$.
(b) For n odd, $x^n + y^n = (x + y)(x^{n-1} - x^{n-2}y + x^{n-3}y^2 - \cdots + y^{n-1})$.

25 If $a^n - 1$ is a prime, show that $a = 2$ and that n is a prime. Primes of the form $2^p - 1$ are called Mersenne primes. For example, $2^3 - 1 = 7$ and $2^5 - 1 = 31$. It is not known if there are infinitely many Mersenne primes.

26 If $a^n + 1$ is a prime, show that a is even and that n is a power of 2. Primes of the form $2^{2^t} + 1$ are called Fermat primes. For example, $2^{2^1} + 1 = 5$ and $2^{2^2} + 1 = 17$. It is not known if there are infinitely many Fermat primes.

27 For all odd n show that $8 \mid n^2 - 1$. If $3 \nmid n$, show that $6 \mid n^2 - 1$.

28 For all n show that $30 \mid n^5 - n$ and that $42 \mid n^7 - n$.

29 Suppose that $a, b, c, d \subseteq \mathbb{Z}$ and that $(a, b) = (c, d) = 1$. If $(a/b) + (c/d) =$ an integer, show that $b = \pm d$.

30 Prove that $\frac{1}{2} + \frac{1}{3} + \cdots + 1/n$ is not an integer.

31 Show that 2 is divisible by $(1 + i)^2$ in $\mathbb{Z}[i]$.

32 For $\alpha = a + bi \in \mathbb{Z}[i]$ we defined $\lambda(\alpha) = a^2 + b^2$. From the properties of λ deduce the identity $(a^2 + b^2)(c^2 + d^2) = (ac - bd)^2 + (ad + bc)^2$.

33 Show that $\alpha \in \mathbb{Z}[i]$ is a unit iff $\lambda(\alpha) = 1$. Deduce that $1, -1, i$, and $-i$ are the only units in $\mathbb{Z}[i]$.

34 Show that 3 is divisible by $(1 - \omega)^2$ in $\mathbb{Z}[\omega]$.

35 For $\alpha = a + b\omega \in \mathbb{Z}[\omega]$ we defined $\lambda(\alpha) = a^2 - ab + b^2$. Show that α is a unit iff $\lambda(\alpha) = 1$. Deduce that $1, -1, \omega, -\omega, \omega^2$, and $-\omega^2$ are the only units in $\mathbb{Z}[\omega]$.

36 Define $\mathbb{Z}[\sqrt{-2}]$ as the set of all complex numbers of the form $a + b\sqrt{-2}$, where $a, b \in \mathbb{Z}$. Show that $\mathbb{Z}[\sqrt{-2}]$ is a ring. Define $\lambda(\alpha) = a^2 + 2b^2$ for $\alpha = a + b\sqrt{-2}$. Use λ to show that $\mathbb{Z}[\sqrt{-2}]$ is a Euclidean domain.

37 Show that the only units in $\mathbb{Z}[\sqrt{-2}]$ are 1 and -1.

38 Suppose that $\pi \in \mathbb{Z}[i]$ and that $\lambda(\pi) = p$ is a prime in $\mathbb{Z}$. Show that π is a prime in $\mathbb{Z}[i]$. Show that the corresponding result holds in $\mathbb{Z}[\omega]$ and $\mathbb{Z}[\sqrt{-2}]$.

39 Show that in any integral domain a prime element is irreducible.

chapter two/APPLICATIONS OF UNIQUE FACTORIZATION

The importance of the notion of prime number should be evident from the results of Chapter 1.

In this chapter we shall give several proofs of the fact that there are infinitely many primes in $\mathbb{Z}$. *We shall also consider the analogous question for the ring* $k[x]$.

The theorem of unique prime decomposition is sometimes referred to as the fundamental theorem of arithmetic. We shall begin to demonstrate its usefulness by using it to investigate the properties of some natural number-theoretic functions.

I *INFINITELY MANY PRIMES IN* $\mathbb{Z}$

Theorem (Euclid)
In the ring $\mathbb{Z}$ *there are infinitely many prime numbers.*

PROOF
Let us consider positive primes. Label them in increasing order $p_1, p_2, p_3, \ldots$. Thus $p_1 = 2, p_2 = 3, p_3 = 5$, etc. Let $N = (p_1 p_2 \cdots p_n) + 1$. N is greater than 1 and not divisible by any p_i, $i = 1, 2, \ldots, n$. On the other hand, N is divisible by some prime, p, and p must be greater than p_n.

We have shown that given any positive prime there is another prime that is greater. It follows that the set of primes is infinite.

The analogous theorem for $k[x]$ is that there are infinitely many monic, irreducible polynomials. If k is infinite, this is trivial since $x - a$ is monic and irreducible for all $a \in k$. This proof does not work if k is finite, but Euclid's proof may easily be adapted to this case. We leave this as an exercise.

Recall that in an integral domain two elements are called associate if they differ only by multiplication by a unit. We now know that in $\mathbb{Z}$ and $k[x]$ there are infinitely many nonassociate primes. It is instructive to consider a ring where all primes are associate, so that in essence there is only one prime.

Let $p \in \mathbb{Z}$ be a prime number and let $\mathbb{Z}_p$ be the set of all rational members a/b, where $p \not| b$. One easily checks using the remark following Corollary 1 to Proposition 1.1.1 that $\mathbb{Z}_p$ is a ring. $a/b \in Z_p$ is a unit if there is a $c/d \in \mathbb{Z}_p$ such that $a/b \cdot c/d = 1$. Then $ac = bd$, which implies $p \not| a$ since $p \not| b$ and $p \not| d$. Conversely, any rational member a/b is a unit in $\mathbb{Z}_p$ if $p \not| a$ and $p \not| b$. If $a/b \in \mathbb{Z}_p$, write $a = p^l a'$, where $p \not| a'$. Then $a/b = p^l a'/b$. Thus every element of $\mathbb{Z}_p$ is a power of p times a unit. From this it is easy to see that the only primes in $\mathbb{Z}_p$ have the form pc/d, where c/d is a unit. Thus all the primes of $\mathbb{Z}_p$ are associate.

Exercise

If $a/b \in \mathbb{Z}_p$ is not a unit, prove that $a/b + 1$ is a unit. This phenomenon shows why Euclid's proof breaks down in general for integral domains.

2 *SOME ARITHMETIC FUNCTIONS*

In the remainder of this chapter we shall give some applications of the unique factorization theorem.

An integer $a \in \mathbb{Z}$ is said to be square-free if it is not divisible by the square of any other integer greater than 1.

Proposition 2.2.1

If $n \in \mathbb{Z}$, n can be written in the form $n = ab^2$, where $a, b \in \mathbb{Z}$ and a is square-free.

PROOF

Let $n = p_1^{a_1} p_2^{a_2} \cdots p_l^{a_l}$. One can write $a_i = 2b_i + r_i$, where $r_i = 0$ or 1 depending on whether a_i is even or odd. Set $a = p_1^{r_1} p_2^{r_2} \cdots p_l^{r_l}$ and $b = p_1^{b_1} p_2^{b_2} \cdots p_l^{b_l}$. Then $n = ab^2$ and a is clearly square-free.

This lemma can be used to give another proof that there are infinitely many primes in $\mathbb{Z}$. Assume that there are not, and let $p_1, p_2, \ldots, p_l$ be a complete list of positive primes. Consider the set of positive integers less than or equal to N. If $n \leq N$, then $n = ab^2$, where a is square-free and thus equal to one of the 2^l numbers $p_1^{\varepsilon_1} p_2^{\varepsilon_2} \cdots p_l^{\varepsilon_l}$, where $\varepsilon_i = 0$ or $1, i = 1, \ldots, l$. Notice that $b \leq \sqrt{N}$. There are at most $2^l \sqrt{N}$ numbers satisfying these conditions and so $N \leq 2^l \sqrt{N}$, or $\sqrt{N} \leq 2^l$, which is clearly false for N large enough. This contradiction proves the result.

It is possible to give a similar proof that there are infinitely many monic irreducibles in $k[x]$, where k is a finite field.

There are a number of naturally defined functions on the integers. For example, given a positive integer n let $\nu(n)$ be the number of positive divisors of n and $\sigma(n)$ the sum of the positive divisors of n. For example, $\nu(3) = 2$, $\nu(6) = 4$, and $\nu(12) = 6$ and $\sigma(3) = 4$, $\sigma(6) = 12$, and $\sigma(12) = 28$. Using unique factorization it is possible to obtain rather simple formulas for these functions.

Proposition 2.2.2

If n is a positive integer, let $n = p_1^{a_1} p_2^{a_2} \cdots p_l^{a_l}$ be its prime decomposition. Then

(a) $\nu(n) = (a_1 + 1)(a_2 + 1) \cdots (a_l + 1)$.
(b) $\sigma(n) = ((p_1^{a_1+1} - 1)/(p_1 - 1))((p_2^{a_2+1} - 1)/(p_2 - 1)) \cdots$
$((p_l^{a_l+1} - 1)/(p_l - 1))$.

PROOF

To prove part (a) notice that $m \mid n$ iff $m = p_1^{b_1} p_2^{b_2} \cdots p_l^{b_l}$ and $0 \leq b_i \leq a_i$ for $i = 1, 2, \ldots, l$. Thus the positive divisors of n are one-to-one correspondence with the n-tuples $(b_1, b_2, \ldots, b_l)$ with $0 \leq b_i \leq a_i$ for $i = 1, \ldots, l$, and there are exactly $(a_1 + 1)(a_2 + 1) \cdots (a_l + 1)$ such n-tuples.

To prove part (b) notice that $\sigma(n) = \sum p_1^{b_1} p_2^{b_2} \cdots p_l^{b_l}$, where the sum is over the above set of n-tuples. Thus, $\sigma(n) = (\sum_{b_1=0}^{a_1} p_1^{b_1})(\sum_{b_2=0}^{a_2} p_2^{b_2}) \cdots (\sum_{b_l=0}^{a_l} p_l^{b_l})$, from which the result follows by use of the summation formula for the geometric series.

There is an interesting and unsolved problem connected with the function $\sigma(n)$. A number n is said to be perfect if $\sigma(n) = 2n$. 6 and 28 are examples. In general, if $2^{m+1} - 1$ is a prime, then $n = 2^m(2^{m+1} - 1)$ is perfect, as can be seen by applying part (b) of Proposition 2.2.2. This fact is already in Euclid. L. Euler showed that any even perfect number has this form. Thus the problem of even perfect numbers is reduced to that of finding primes of the form $2^{m+1} - 1$. Such primes are called Mersenne primes. The two outstanding problems involving perfect numbers are the following: Are there infinitely many perfect numbers? Are there any odd perfect numbers?

The multiplicative analog of this problem is trivial. An integer n is called multiplicatively perfect if the product of the positive divisors of n is n^2. Such a number cannot be a prime or a square of a prime. Thus there is a proper divisor d such that $d \neq n/d$. The product of the divisors $1, d, n/d$, and n is already n^2. Thus n is multiplicatively perfect iff there are exactly two proper divisors. The only such numbers are cubes of primes or products of two distinct primes. For example, 27 and 10 are multiplicatively perfect.

We now introduce a very important arithmetic function, the Möbius μ function. For $n \in \mathbb{Z}^+$, $\mu(1) = 1$, $\mu(n) = 0$ if n is not square-free, and $\mu(p_1 p_2 \cdots p_l) = (-1)^l$, where the p_i are distinct positive primes.

Proposition 2.2.3
If $n > 1$, $\sum_{d|n} \mu(d) = 0$.

PROOF
If $n = p_1^{a_1} p_2^{a_2} \cdots p_l^{a_l}$, then $\sum_{d|n} \mu(d) = \sum_{(\varepsilon_1, \ldots, \varepsilon_l)} \mu(p_1^{\varepsilon_1} \cdots p_l^{\varepsilon_l})$, where the ε_i are zero or 1. Thus $\sum_{d|n} \mu(d) = 1 - l + \binom{l}{2} - \binom{l}{3} + \cdots + (-1)^l = (1 - 1)^l = 0$.

The full significance of the Möbius μ function can be understood most clearly when its connection with Dirichlet multiplication is brought to light. Let f and g be complex valued functions on $\mathbb{Z}^+$. The Dirichlet product of f and g is defined by the formula $f \circ g(n) = \sum f(d_1) g(d_2)$, where the sum is over all pairs (d_1, d_2) of positive integers such that $d_1 d_2 = n$. This product is associative, as one can see by checking that $f \circ (g \circ h)(n) = (f \circ g) \circ h(n) = \sum f(d_1) g(d_2) h(d_3)$, where the sum is over all 3-tuples (d_1, d_2, d_3) of positive integers such that $d_1 d_2 d_3 = n$.

Define the function $\mathbb{I}$ by $\mathbb{I}(1) = 1$ and $\mathbb{I}(n) = 0$ for $n > 1$. Then $f \circ \mathbb{I} = \mathbb{I} \circ f = f$. Define I by $I(n) = 1$ for all $n \in Z^+$. Then $f \circ I(n) = I \circ f(n) = \sum_{d|n} f(d)$.

Lemma
$I \circ \mu = \mu \circ I = \mathbb{I}$.

PROOF
$\mu \circ I(1) = \mu(1) I(1) = 1$. If $n > 1$, $\mu \circ I(n) = \sum_{d|n} \mu(d) = 0$. The same proof works for $I \circ \mu$.

Theorem 1 (Möbius Inversion Theorem)
Let $F(n) = \sum_{d|n} f(d)$. *Then* $f(n) = \sum_{d|n} \mu(d) F(n/d)$.

PROOF
$F = f \circ I$. Thus $F \circ \mu = (f \circ I) \circ \mu = f \circ (I \circ \mu) = f \circ \mathbb{I} = f$. This shows that $f(n) = F \circ \mu(n) = \sum_{d|n} \mu(d) F(n/d)$.

Remark
We have considered complex-valued functions on the positive integers. It is useful to notice that Theorem 1 is valid whenever the functions

take their value in an abelian group. The proof goes through word for word.

If the group law in the abelian group is written multiplicatively, the theorem takes the following form: If $F(n) = \prod_{d|n} f(d)$, then $f(n) = \prod_{d|n} F(n/d)^{\mu(d)}$.

The Möbius inversion theorem has many applications. We shall use it to obtain a formula for yet another arithmetic function, the Euler ϕ function. For $n \in Z^+$, $\phi(n)$ is defined to be the number of integers between 1 and n relatively prime to n. For example, $\phi(1) = 1$, $\phi(5) = 4$, $\phi(6) = 2$, and $\phi(9) = 6$. If p is a prime, it is clear that $\phi(p) = p - 1$.

Proposition 2.2.4

$\sum_{d|n} \phi(d) = n$.

PROOF

Consider the n rational numbers $1/n, 2/n, 3/n, \ldots, (n-1)/n, n/n$. Reduce each to lowest terms; i.e., express each number as a quotient of relatively prime integers. The denominators will all be divisors of n. If $d \mid n$, exactly $\phi(d)$ of our numbers will have d in the denominator after reducing to lowest terms. Thus $\sum_{d|n} \phi(d) = n$.

Proposition 2.2.5

If $n = p_1^{a_1} p_2^{a_2} \cdots p_l^{a_l}$, *then* $\phi(n) = n(1 - (1/p_1))(1 - (1/p_2)) \cdots (1 - (1/p_l))$.

PROOF

Since $n = \sum_{d|n} \phi(d)$ the Möbius inversion theorem implies that $\phi(n) = \sum_{d|n} \mu(d) n/d = n - \sum_i n/p_i + \sum_{i \neq j} n/p_i p_j \cdots = n(1 - (1/p_1))(1 - (1/p_2)) \cdots (1 - (1/p_l))$.

Later we shall give a more insightful proof of this formula. We shall also use the Möbius function to determine the number of monic irreducible polynomials of fixed degree in $k[x]$, where k is a finite field.

3 $\sum 1/p$ *DIVERGES*

We began this chapter by proving that there are infinitely many prime numbers in $\mathbb{Z}$. We shall conclude by proving a somewhat stronger statement. The proof will assume some elementary facts from the theory of infinite series.

Theorem 2

$\sum 1/p$ diverges, where the sum is over all positive primes in $\mathbb{Z}$.

PROOF

Let $p_1, p_2, \ldots, p_{l(n)}$ be all the primes less than n and define $\lambda(n) = \prod_{i=1}^{l(n)} (1 - 1/p_i)^{-1}$. Since $(1 - 1/p_i)^{-1} = \sum_{a_i=1}^{\infty} 1/p_i^{a_i}$ we see that

$$\lambda(n) = \sum (p_1^{a_1} p_2^{a_2} \cdots p_l^{a_l})^{-1}$$

where the sum is over all l-tuples of nonnegative integers $(a_1, a_2, \ldots, a_l)$. In particular, we see that $1 + \frac{1}{2} + \frac{1}{3} + \cdots + 1/n < \lambda(n)$. Thus $\lambda(n) \to \infty$ as $n \to \infty$. This already gives a new proof that there are finitely many primes.

Next, consider $\log \lambda(n)$. We have

$$\log \lambda(n) = -\sum_{i=1}^{l} \log(1 - p_i^{-1}) = \sum_{i=1}^{l} \sum_{m=1}^{\infty} (mp_i^m)^{-1}$$

$$= p_1^{-1} + p_2^{-1} + \cdots + p_l^{-1} + \sum_{i=1}^{l} \sum_{m=2}^{\infty} (mp_i^m)^{-1}$$

Now, $\sum_{m=2}^{\infty} (mp_i^m)^{-1} < \sum_{m=2}^{\infty} p_i^{-m} = p_i^{-2}(1 - p_i^{-1})^{-1} \leq 2p_i^{-2}$. Thus $\log \lambda(n) < p_1^{-1} + p_2^{-1} + \cdots + p_l^{-1} + 2(p_1^{-2} + p_2^{-2} + \cdots + p_l^{-2})$. It is well known that $\sum_{n=1}^{\infty} n^{-2}$ converges. It follows that $\sum_{i=1}^{\infty} p_i^{-2}$ converges. Thus if $\sum p^{-1}$ converged, there would be a constant M such that $\log \lambda(n) < M$, or $\lambda(n) < e^M$. This, however, is impossible since $\lambda(n) \to \infty$ as $n \to \infty$. Thus $\sum p^{-1}$ diverges.

It is instructive to try to construct an analog of Theorem 2 for the ring $k[x]$, where k is a finite field with q elements. The role of the positive primes p is taken by the monic irreducible polynomials $p(x)$. The "size" of a monic polynomial $f(x)$ is given by the quantity $q^{\deg f(x)}$.

This is reasonable because for a positive integer n, n is the number of nonnegative integers less than n, i.e., the number of elements in the set $\{0, 1, 2, \ldots, n - 1\}$. Analogously, $q^{\deg f(x)}$ is the number of polynomials of degree less than $\deg f(x)$. This is easy to see. Any such polynomial has the form $a_0 x^m + a_1 x^{m-1} + \cdots + a_m$, where $m = \deg f(x) - 1$ and $a_i \in k$. There are q choices for a_i and the choice for each index is independent of the others. Thus there are $q^{m+1} = q^{\deg f(x)}$ such polynomials.

Theorem 3

$\sum q^{-\deg p(x)}$ diverges, where the sum is over all monic irreducibles $p(x)$ in $k[x]$.

PROOF

We first show that $\sum q^{-\deg f(x)}$ diverges and that $\sum q^{-2\deg f(x)}$ converges, where both sums are over all monic polynomials $f(x)$ in $k[x]$. Both results follow from the fact that there are exactly q^n monic polynomials of degree n in $k[x]$. Consider $\sum_{\deg f(x) \leq n} q^{-\deg f(x)}$. This sum is equal to $\sum_{m=0}^{n} q^m q^{-m} = n + 1$. Thus $\sum q^{-\deg f(x)}$ diverges. Similarly, $\sum_{\deg f(x) \leq n} q^{-2\deg f(x)} = \sum_{m=0}^{n} q^m q^{-2m} < (1 - 1/q)^{-1}$. Thus $\sum q^{-2\deg f(x)}$ converges.

The rest of the proof is an exact imitation of the proof of Theorem 2. The reader should fill in the details.

Notes

There are a multitude of unsolved problems in the theory of prime numbers. We have given some examples previously. A famous theorem due to L. Dirichlet states that there are infinitely many prime numbers of the form $ax + b$, where a and b are relatively prime integers. A number of special cases are proved in this book. For a proof of the full result see [64]. By contrast it is unknown if there are infinitely many primes of the form $x^2 + 1$, or indeed any polynomial function of x. G. H. Hardy and J. E. Littlewood have conjectured that there are infinitely many primes of the form $x^2 + 1$ and have even conjectured a formula for the number of such primes lying below a given bound (see D. Shanks [70]). We shall show that a prime divides a number of the form $x^2 + 1$ iff the prime is of the form $4t + 1$. This and other results like it follow from the law of quadratic reciprocity and its generalizations.

Good discussions of unsolved problems about primes are given in W. Sierpinski [71] and Shanks [70]. For readers with a more extensive background in analysis, see the paper by P. Erdös [31] and those of Hardy [38] and [39].

The key idea behind the proof of Theorem 2 goes back to Euler, who observed that

$$\prod_{p \leq n} \left(1 - \frac{1}{p}\right)^{-1} > 1 + \frac{1}{2} + \cdots + \frac{1}{n}$$

A pleasant account of this for the beginner can be found in Rademacher and Toeplitz [65].

Theorem 3 is a Euler-style proof of the existence of infinitely many irreducibles in $k[x]$, where k is a finite field. This suggests that many of the theorems in classical number theory have analogs in the ring $k[x]$. This is indeed the case. An interesting reference along these lines is L. Carlitz [10]. The Dirichlet theorem about primes in an arithmetic progression, mentioned above, has been proved for $k[x]$ by H. Kornblum [50]. (Kornblum had his promising career cut short after he enlisted as a Kriegsfreiwilliger in 1914.) The prime number theorem, which we mentioned in Chapter 1, also has its analog in $k[x]$. This was proved by E. Artin in his doctoral thesis [2].

Exercises

1 Show that $k[x]$, with k a finite field, has infinitely many irreducible polynomials.

2 Let $p_1, p_2, \ldots, p_t \in \mathbb{Z}$ be primes and consider the set of all rational numbers $r = a/b$, $a, b \in \mathbb{Z}$, such that $\text{ord}_{p_i} a \geq \text{ord}_{p_i} b$ for $i = 1, 2, \ldots, t$. Show that this set is a ring and that up to taking associates $p_1, p_2, \ldots, p_t$ are the only primes.

3 Use the formula for $\varphi(n)$ to give a proof that there are infinitely many primes. [*Hint:* If $p_1, p_2, \ldots, p_t$ were all the primes, then $\varphi(n) = 1$, where $n = p_1 p_2 \cdots p_t$.]

4 If a is a nonzero integer, then for $n > m$ show that $(a^{2^n} + 1, a^{2^m} + 1) = 1$ or 2 depending on whether a is odd or even. (*Hint:* If p is an odd prime and $p \mid a^{2^m} + 1$, then $p \mid a^{2^n} - 1$ for $n > m$.)

5 Use the result of Exercise 4 to show that there are infinitely many primes. (This proof is due to G. Polya.)

6 For a rational number r let $[r]$ be the largest integer less than or equal to r, e.g., $[\frac{1}{2}] = 0$, $[2] = 2$, and $[3\frac{1}{3}] = 3$. Prove $\text{ord}_p n! = [n/p] + [n/p^2] + [n/p^3] + \cdots$.

7 Deduce from Exercise 6 that $\text{ord}_p n! \leq n/(p-1)$ and that $\sqrt[n]{n!} \leq \prod_{p \mid n!} p^{1/(p-1)}$.

8 Use Exercise 7 to show that there are infinitely many primes. [*Hint:* $(n!)^2 \geq n^n$.] (This proof is due to Eckford Cohen.)

9 A function on the integers is said to be multiplicative if $f(ab) = f(a)f(b)$ whenever $(a, b) = 1$. Show that a multiplicative function is completely determined by its value on prime powers.

10 If $f(n)$ is a multiplicative function, show that the function $g(n) = \sum_{d \mid n} f(d)$ is also multiplicative.

11 Show that $\varphi(n) = n \sum_{d \mid n} \mu(d)/d$ by first proving that $\mu(d)/d$ is multiplicative and then using Exercises 9 and 10.

12 Find formulas for $\sum_{d \mid n} \mu(d)\varphi(d)$, $\sum_{d \mid n} \mu(d)^2 \varphi(d)^2$, and $\sum_{d \mid n} \mu(d)/\varphi(d)$.

13 Let $\sigma_k(n) = \sum_{d \mid n} d^k$. Show that $\sigma_k(n)$ is multiplicative and find a formula for it.

14 If $f(n)$ is multiplicative, show that $h(n) = \sum_{d \mid n} \mu(n/d) f(d)$ is also multiplicative.

15 Show that
(a) $\sum_{d \mid n} \mu(n/d)\nu(d) = 1$ for all n.
(b) $\sum_{d \mid n} \mu(n/d)\sigma(d) = n$ for all n.

16 Show that $\nu(n)$ is odd iff n is a square.

17 Show that $\sigma(n)$ is odd iff n is a square or twice a square.

18 Prove that $\varphi(n)\varphi(m) = \varphi((n, m))\varphi([n, m])$.

19 Prove that $\varphi(mn)\varphi((m, n)) = (m, n)\varphi(m)\varphi(n)$.

20 Prove that $\prod_{d|n} d = n^{\nu(n)/2}$.

21 Define $\wedge(n) = \log p$ if n is a power of p and zero otherwise. Prove that $\sum_{d|n} \mu(n/d) \log d = \wedge(n)$. [*Hint:* First calculate $\sum_{d|n} \wedge(d)$ and then apply the Möbius inversion formula.]

22 Show that the sum of all the integers t such that $1 \le t \le n$ and $(t, n) = 1$ is $\frac{1}{2}n\varphi(n)$.

23 Let $f(x) \in \mathbb{Z}[x]$ and let $\psi(n)$ be the number of $f(j)$, $j = 1, 2, \ldots, n$, such that $(f(j), n) = 1$. Show that $\psi(n)$ is multiplicative and that $\psi(p^t) = p^{t-1}\psi(p)$. Conclude that $\psi(n) = n \prod_{p|n} \psi(p)/p$.

24 Supply the details to the proof of Theorem 3.

25 Consider the function $\zeta(s) = \sum_{n=1}^{\infty} 1/n^s$. $\zeta(s)$ is called the Riemann zeta function. It converges for $s > 1$. Prove the formal identity (Euler's identity) $\zeta(s) = \prod_p (1 - (1/p^s))^{-1}$. If we let s assume complex values, it can be shown that $\zeta(s)$ has an analytic continuation to the whole complex plane. The famous Riemann hypothesis states that the only zeros of $\zeta(s)$ lying in the strip $0 \le \operatorname{Re} s \le 1$ lie on the line $\operatorname{Re} s = \frac{1}{2}$.

26 Verify the formal identities

(a) $\zeta(s)^{-1} = \sum_{n=1}^{\infty} \mu(n)/n^s$.

(b) $\zeta(s)^2 = \sum_{n=1}^{\infty} \nu(n)/n^s$.

(c) $\zeta(s)\zeta(s - 1) = \sum_{n=1}^{\infty} \sigma(n)/n^s$.

27 Show that $\sum' 1/n$, the sum being over square free integers, diverges. Conclude that $\prod_{p<N} (1 + 1/p) \to \infty$ as $N \to \infty$. Since $e^x > 1 + x$, conclude that $\sum_{p<N} 1/p \to \infty$. (This proof is due to I. Niven.)

chapter three/CONGRUENCE

Gauss first introduced the notion of congruence in Disquisitiones Arithmeticae (*see Notes in Chapter* 1). *It is an extremely simple idea. Nevertheless, its importance and usefulness in number theory cannot be exaggerated.*

This chapter is devoted to an exposition of the simplest properties of congruence. In Chapter 4, *we shall go into the subject in more depth.*

I *ELEMENTARY OBSERVATIONS*

It is a simple observation that the product of two odd numbers is odd, the product of two even numbers is even, and the product of an odd and even number is even. Also, notice that an odd plus an odd is even, an even plus an even is even, and an even plus an odd is odd. This information is summarized in Tables 1 and 2. Table 1 is like a multiplication table and Table 2 like an addition table.

Table 1

	e	o
e	e	e
o	e	o

Table 2

	e	o
e	e	o
o	o	e

These observations are so elementary one might ask if anything interesting can be deduced from them. The answer, surprisingly, is yes.

Many problems in number theory have the form, If f is a polynomial in one or several variables with integer coefficients, does the equation $f = 0$ have integer solutions? Such questions were considered by the Greek mathematician Diophantus and are called Diophantine problems in his honor.

Consider the equation $x^2 - 117x + 31 = 0$. We claim that there is no solution that is an integer. Let n be any integer. n is either even or odd. If n is even, so is n^2 and $117n$. Thus $n^2 - 117n + 31$ is odd. If n is odd, then n^2 and $117n$ are both odd. Thus $n^2 - 117n + 31$ is odd in this case also. Since every integer is even or odd, this shows that $n^2 - 117n + 31$ is never zero.

In Chapter 2 we showed that there are infinitely many prime numbers. We shall now show that there are infinitely many prime numbers that leave a remainder of 3 when divided by 4. Examples of such primes are 3, 7, 19, and 59.

An integer divided by 4 leaves a remainder of 0, 1, 2, or 3. Thus odd numbers are either of the form $4k + 1$ or $4l + 3$. The product of two numbers of the form $4k + 1$ is again of that form; $(4k + 1)(4k' + 1) = 4(4kk + k + k') + 1$. It follows that an integer of the form $4l + 3$ must be divisible by a prime of the form $4l + 3$.

Now, suppose that there were only finitely many positive primes of the form $4l + 3$. This list begins 3, 7, 11, 19, 23, ... Let $p_1 = 7$, $p_2 = 11$, $p_3 = 19$, etc. Suppose that p_m is the largest prime of this form and set $N = 4p_1p_2 \cdots p_m + 3$. N is not divisible by any of the p_i. However, N is of the form $4l + 3$ and so must be divisible by a prime p of the form $4l + 3$. We have $p > p_m$, which is a contradiction.

There is clearly some common principle underlying both arguments. We explore this in Section 2.

2 CONGRUENCE IN $\mathbb{Z}$

Definition

If $a, b, m \in \mathbb{Z}$ and $m \neq 0$, we say that a is *congruent to b modulo m* if m divides $b - a$. This relation is written $a \equiv b\ (m)$.

Proposition 3.2.1

(a) $a \equiv a\ (m)$.
(b) $a \equiv b\ (m)$ *implies that* $b \equiv a\ (m)$.
(c) *If* $a \equiv b\ (m)$ *and* $b \equiv c\ (m)$, *then* $a \equiv c\ (m)$.

PROOF

(a) $a - a = 0$ and $m \mid 0$.
(b) If $m \mid b - a$, then $m \mid a - b$.
(c) If $m \mid b - a$ and $m \mid c - b$, then $m \mid c - a = (c - b) + (b - a)$.

Proposition 3.2.1 shows that congruence modulo m is an equivalence relation on the set of integers. If $a \in \mathbb{Z}$, let $\bar{a}$ denote the set of integers congruent to a modulo m. $\bar{a} = \{n \in \mathbb{Z} \mid n \equiv a\ (m)\}$, or in other words $\bar{a}$ is the set of integers of the form $a + km$.

If $m = 2$, then $\bar{0}$ is the set of even integers and $\bar{1}$ is the set of odd integers.

Definition

A set of the form $\bar{a}$ is called a *congruence class modulo m*.

Proposition 3.2.2

(a) $\bar{a} = \bar{b}$ *iff* $a \equiv b\,(m)$.

(b) $\bar{a} \neq \bar{b}$ *iff* $\bar{a} \cap \bar{b}$ *is empty.*

(c) *There are precisely m distinct congruence classes modulo m.*

PROOF

(a) If $\bar{b} = \bar{a}$, then $a \in \bar{a} = \bar{b}$. Thus $a \equiv b\,(m)$. Conversely, if $a \equiv b\,(m)$, then $a \in \bar{b}$. If $c \equiv a\,(m)$, then $c \equiv b\,(m)$, which shows $\bar{a} \subseteq \bar{b}$. Since $a \equiv b\,(m)$ implies that $b \equiv a\,(m)$, we also have $\bar{b} \subseteq \bar{a}$. Therefore $\bar{a} = \bar{b}$.

(b) Clearly, if $\bar{a} \cap \bar{b}$ is empty, then $\bar{a} \neq \bar{b}$. We shall show that $\bar{a} \cap \bar{b}$ not empty implies that $\bar{a} = \bar{b}$. Let $c \in \bar{a} \cap \bar{b}$. Then $c \equiv a\ (m)$ and $c \equiv b\ (m)$. It follows that $a \equiv b\ (m)$ and so by part (a) we have $\bar{a} = \bar{b}$.

(c) We shall show that $\bar{0}, \bar{1}, \bar{2}, \ldots, \overline{m - 1}$ are all distinct and are a complete set of congruence classes modulo m. Suppose that $0 \leq k < l < m$. $\bar{k} = \bar{l}$ implies that $k \equiv l\ (m)$ or that m divides $l - k$. Since $0 < l - k < m$ this is a contradiction. Therefore $\bar{k} \neq \bar{l}$. Now let $a \in \mathbb{Z}$. We can find integers q and r such that $a = qm + r$, where $0 \leq r < m$. It follows that $a \equiv r\ (m)$ and that $\bar{a} = \bar{r}$.

Definition

The set of congruence classes modulo m is denoted by $\mathbb{Z}/m\mathbb{Z}$.

If $\bar{a}_1, \bar{a}_2, \ldots, \bar{a}_m$ are a complete set of congruence classes modulo m, then $\{a_1, a_2, \ldots, a_m\}$ is called a *complete set of residues modulo m.*

For example, $\{0, 1, 2, 3\}$, $\{4, 9, 14, -1\}$, and $\{0, 1, -2, -1\}$ are complete sets of residues modulo 4.

The set $\mathbb{Z}/m\mathbb{Z}$ can be made into a ring by defining in a natural way addition and multiplication. This is accomplished by means of the following proposition.

Proposition 3.2.3

If $a \equiv c\ (m)$ and $b \equiv d\ (m)$, then $a + b \equiv c + d\ (m)$ and $ab \equiv cd\ (m)$.

PROOF

If $m \mid c - a$ and $m \mid d - b$, then $m \mid (c - a) + (d - b) = (c + d) - (a + b)$. Thus $a + b \equiv c + d\ (m)$.

Notice that $cd - ab = c(d - b) + b(c - a)$. Thus $m \mid cd - ab$ and $ab \equiv cd\ (m)$.

If $\bar{a}, \bar{b} \in \mathbb{Z}/m\mathbb{Z}$, we define $\bar{a} + \bar{b}$ to be $\overline{a + b}$ and $\bar{a}\bar{b}$ to be $\overline{ab}$.

This definition seems to depend on a and b. We have to show that they depend only on the congruence classes defined by a and b. This is easy. Assume that $\bar{c} = \bar{a}$ and that $\bar{d} = \bar{b}$. We must show that $\overline{a + b} =$

$\overline{c + d}$ and that $\overline{ab} = \overline{cd}$, but this follows immediately from Propositions 3.2.2 and 3.2.3.

With these definitions $\mathbb{Z}/m\mathbb{Z}$ becomes a ring. The verification of this fact is left to the reader.

Tables 3 and 4 give explicitly the addition and multiplication in $\mathbb{Z}/3\mathbb{Z}$. (Bars over the numbers are omitted.) The reader should construct similar tables for $m = 4$, 5, and 6.

Table 3
Addition

	0	1	2
0	0	1	2
1	1	2	0
2	2	0	1

Table 4
Multiplication

	0	1	2
0	0	0	0
1	0	1	2
2	0	2	1

In discussing arithmetic problems it is sometimes more convenient to work with the ring $\mathbb{Z}/m\mathbb{Z}$ than with the notion of congruence modulo m. On the other hand, it is sometimes more convenient the other way around. We shall switch back and forth between the two viewpoints as the situation demands.

We proved earlier that the polynomial $x^2 - 117x + 31$ has no integer roots. It is possible to generalize this result using some of the material we have developed.

If $a \equiv b\ (m)$, then $a^2 \equiv b^2\ (m)$, $a^3 \equiv b^3\ (m)$, and in general $a^n \equiv b^n\ (m)$. It follows that if $p(x) \in \mathbb{Z}[x]$, then $p(a) \equiv p(b)\ (m)$. All this is a consequence of Proposition 3.2.3.

Take $m = 2$. Then a is congruent to either 0 or 1 modulo 2 and we have $p(a) \equiv p(0)\ (2)$ or $p(a) \equiv p(1)\ (2)$.

If $p(x) = a_0x^n + a_1x^{n-1} + \cdots + a_{n-1}x + a_n$, then $p(0) = a_n$ and $p(1) = a_0 + a_1 + \cdots + a_n$. Our calculations yield the following result: If $p(x) \in \mathbb{Z}[x]$ and $p(0)$ and $p(1)$ are both odd, then $p(x)$ has no integer roots.

$x^2 - 117x + 31$ has constant term 31, and the sum of the coefficients is -85, both of which are odd. Other examples are $2x^2 + 2x + 1$ and $3x^3 + 2x^2 + x + 3$.

3 *THE CONGRUENCE* $ax \equiv b\ (m)$

The simplest congruence is $ax \equiv b\ (m)$. In this section we shall develop a criterion to test this congruence for solvability, and if it is solvable, give a formula for the number of solutions.

Before beginning we must give a definition of what we mean by the number of solutions to a congruence. Quite generally, let $f(x_1, \ldots, x_n)$ be a polynomial in n variables with integer coefficients and consider the congruence $f(x_1, \ldots, x_n) \equiv 0\ (m)$. A solution is an n-tuple of integers $(a_1, \ldots, a_n)$ such that $f(a_1, a_2, \ldots, a_n) \equiv 0\ (m)$. If $(b_1, \ldots, b_n)$ is another n-tuple such that $b_i \equiv a_i\ (m)$ for $i = 1, \ldots, n$, then it is easy to see that $f(b_1, \ldots, b_n) \equiv 0\ (m)$. We do not want to consider these two solutions as being essentially different. Thus two solutions $(a_1, \ldots, a_n)$ and $(b_1, \ldots, b_n)$ are called equivalent if $a_i \equiv b_i$ for $i = 1, \ldots, n$. The number of solutions to $f(x_1, \ldots, x_n) \equiv 0\ (m)$ is defined to be the number of inequivalent solutions.

For example, 3, 8, and 13 are solutions to $6x \equiv 3\ (15)$. 18 is also a solution, but the solution $x = 18$ is equivalent to the solution $x = 3$.

It is useful to consider the matter from another point of view. The map from $\mathbb{Z}$ to $\mathbb{Z}/m\mathbb{Z}$ given by $a \to \bar{a}$ is a homomorphism. If $f(a_1, \ldots, a_n) \equiv 0\ (m)$, then $\bar{f}(\bar{a}_1, \ldots, \bar{a}_n) = \bar{0}$. Here $\bar{f}(x_1, \ldots, x_n) \in \mathbb{Z}/m\mathbb{Z}[x_1, \ldots, x_n]$ is the polynomial obtained from f by putting a bar over each coefficient of f. One can now see that equivalence classes of solutions to $f(x_1, \ldots, x_n) = 0$ are in one-to-one correspondence with solutions to $\bar{f}(x_1, \ldots, x_n) = \bar{0}$ in the ring $\mathbb{Z}/m\mathbb{Z}$. This interpretation of the number of solutions arises frequently.

We now return to the number of solutions of the congruence $ax \equiv b\ (m)$.

Let $d > 0$ be the greatest common divisor of a and m. Set $a' = a/d$ and $m' = m/d$. Then a' and m' are relatively prime.

Proposition 3.3.1

The congruence $ax \equiv b\ (m)$ has solutions iff $d \mid b$. If $d \mid b$, then there are exactly d solutions. If x_0 is a solution, then the other solutions are given by $x_0 + m', x_0 + 2m', \ldots, x_0 + (d-1)m'$.

PROOF

If x_0 is a solution, then $ax_0 - b = my_0$ for some integer y_0. Thus $ax_0 - my_0 = b$. Since d divides $ax_0 - my_0$, we must have $d \mid b$.

Conversely, suppose that $d \mid b$. By Lemma 4 on page 5 there exist integers x'_0 and y'_0 such that $ax'_0 - my'_0 = d$. Let $c = b/d$ and multiply both sides of the equation by c. Then $a(x'_0c) - m(y'_0c) = b$. Let $x_0 = x'_0c$. Then $ax_0 \equiv b\ (m)$.

We have shown that $ax \equiv b\ (m)$ has a solution iff $d \mid b$.

Suppose that x_0 and x_1 are solutions. $ax_0 \equiv b\ (m)$ and $ax_1 \equiv b\ (m)$ imply that $a(x_1 - x_0) \equiv 0\ (m)$. Thus $m \mid a(x_1 - x_0)$ and $m' \mid a'(x_1 - x_0)$, which implies that $m' \mid x_1 - x_0$ or $x_1 = x_0 + km'$ for some integer k. One easily checks that any number of the form $x_0 + km'$ is a solution and that the solutions $x_0, x_0 + m', \ldots, x_0 + (d-1)m'$ are inequivalent. Let $x_1 = x_0 + km'$ be another solution. There are integers r and s such that $k = rd + s$ and $0 \leq s < d$. Thus $x_1 = x_0 + sm' + rm$ and x_1 is equivalent to $x_0 + sm'$. This completes the proof.

As an example, let us consider the congruence $6x \equiv 3\,(15)$ once more. We first solve $6x - 15y = 3$. Dividing by 3, we have $2x - 5y = 1$. $x = 3$, $y = 1$ is a solution. Thus $x_0 = 3$ is a solution to $6x \equiv 3\ (15)$. Now, $m = 15$ and $d = 3$ so that $m' = 5$. The three inequivalent solutions are 3, 8, and 13.

We have two important corollaries.

Corollary 1

If a and m are relatively prime, then $ax \equiv b\,(m)$ has one and only one solution.

PROOF

In this case $d = 1$ so clearly $d \mid b$, and there are $d = 1$ solutions.

Corollary 2

If p is a prime and $a \not\equiv 0\ (p)$, then $ax \equiv b\ (p)$ has one and only one solution.

PROOF

Immediate from Corollary 1.

Corollaries 1 and 2 can be interpreted in terms of the ring $\mathbb{Z}/m\mathbb{Z}$. The congruence $ax \equiv b\ (m)$ is equivalent to the equation $\bar{a}x = \bar{b}$ over the ring $\mathbb{Z}/m\mathbb{Z}$.

What are the units of $\mathbb{Z}/m\mathbb{Z}$? $\bar{a} \in \mathbb{Z}/m\mathbb{Z}$ is a unit iff $\bar{a}x = \bar{1}$ is solvable. $ax \equiv 1\ (m)$ is solvable iff $d \mid 1$, i.e., iff a and m are relatively prime. Thus $\bar{a}$ is a unit iff $(a, m) = 1$, and it follows easily that there are exactly $\phi(m)$ units in $\mathbb{Z}/m\mathbb{Z}$ [see page 24 for the definition of $\phi(m)$].

If p is a prime and $\bar{a} \neq \bar{0}$ is in $\mathbb{Z}/p\mathbb{Z}$, then $(a, p) = 1$. Thus every nonzero element of $\mathbb{Z}/p\mathbb{Z}$ is a unit, which shows that $\mathbb{Z}/p\mathbb{Z}$ is a field.

If m is not a prime, then $m = m_1 m_2$, where $0 < m_1, m_2 < m$. Thus $\overline{m_1} \neq \bar{0}$, $\overline{m_2} \neq \bar{0}$, but $\overline{m_1}\overline{m_2} = \overline{m_1 m_2} = \bar{m} = \bar{0}$. Therefore $\mathbb{Z}/m\mathbb{Z}$ is not a field.

Summarizing we have

Proposition 3.3.2

An element $\bar{a}$ of $\mathbb{Z}/m\mathbb{Z}$ is a unit iff $(a, m) = 1$. There are exactly $\phi(m)$ units in $\mathbb{Z}/m\mathbb{Z}$. $\mathbb{Z}/m\mathbb{Z}$ is a field iff m is a prime.

***Corollary 1* (*Euler's Theorem*)**
If $(a, m) = 1$, *then* $a^{\phi(m)} \equiv 1\ (m)$.

PROOF
The units in $\mathbb{Z}/m\mathbb{Z}$ form a group of order $\phi(m)$. If $(a, m) = 1$, $\bar{a}$ is a unit. Thus $\bar{a}^{\phi(m)} = \bar{1}$ or $a^{\phi(m)} \equiv 1\ (m)$.

***Corollary 2* (*Fermat's Little Theorem*)**
If p *is a prime and* $p \nmid a$, *then* $a^{p-1} \equiv 1\ (p)$.

PROOF
If $p \nmid a$, then $(a, p) = 1$. Thus $a^{\phi(p)} \equiv 1\ (p)$. The result follows, since for a prime p, $\phi(p) = p - 1$.

It is possible to generalize many of the results in this section to principal ideal domains.

The notions of congruence and residue class can be carried over to an arbitrary commutative ring. The first part of Proposition 3.3.1 is valid in a PID; i.e., $ax \equiv b\ (m)$ has a solution iff $d \mid b$ and the solution is unique iff a and m are relatively prime. The only difference is that the number of solutions need not be finite. In any case, using this result one proves in analogy to part of Proposition 3.3.2 that if R is a PID and $m \in R$ is not zero or a unit, then $R/(m)$ is a field iff m is a prime.

In particular, if k is a field, then $k[x]/(f(x))$ is a field iff $f(x)$ is irreducible.

4 *THE CHINESE REMAINDER THEOREM*

When the modulus m of a congruence is composite it is sometimes possible to reduce a congruence modulo m to a system of simpler congruences. The main theorem of this type is the so-called Chinese remainder theorem (Theorem 1), which we prove below. This theorem is valid for any PID (in fact, even more generally). However, we shall continue to work in $\mathbb{Z}$ and leave to the reader the relatively simple exercise of carrying over the proof for PID's.

Lemma 1
If $a_1, \ldots, a_t$ *are all relatively prime to* m, *then so is* $a_1 a_2 \cdots a_t$.

PROOF
$\bar{a}_i \in \mathbb{Z}/m\mathbb{Z}$ is a unit. Thus so is $\bar{a}_1 \bar{a}_2 \cdots \bar{a}_t = \overline{a_1 a_2 \cdots a_t}$. By Proposition 3.3.2 $a_1 a_2 \cdots a_t$ is relatively prime to m.

Another proof goes as follows. If $a_1a_2 \cdots a_t$ were not prime to m, there would be a prime p that divides them both. $p \mid a_1a_2 \cdots a_t$ implies that $p \mid a_i$ for some i. It follows that $(a_i, m) \neq 1$, which contradicts the hypothesis.

Lemma 2

Suppose that $a_1, \ldots, a_t$ all divide n and that $(a_i, a_j) = 1$ for $i \neq j$. Then $a_1a_2 \cdots a_t$ divides n.

PROOF

The proof is by induction on t. If $t = 1$, there is nothing to do. Suppose that $t > 1$ and that the lemma is true for $t - 1$. Then $a_1a_2 \cdots a_{t-1}$ divides n. By Lemma 1, a_t is prime to $a_1a_2 \cdots a_{t-1}$. Thus there are integers r and s such that $ra_t + sa_1a_2 \cdots a_{t-1} = 1$. Multiply both sides by n. Inspection shows that the left-hand side is divisible by $a_1a_2 \cdots a_t$ and the result follows.

Theorem 1 (Chinese Remainder Theorem)

Suppose that $m = m_1m_2 \cdots m_t$ and that $(m_i, m_j) = 1$ for $i \neq j$. Let $b_1, b_2, \ldots, b_t$ be integers and consider the system of congruences:

$$x \equiv b_1\,(m_1), x \equiv b_2\,(m_2), \ldots, x \equiv b_t\,(m_t)$$

This system always has solutions and any two solutions differ by a multiple of m.

PROOF

Let $n_i = m/m_i$. By Lemma 1, $(m_i, n_i) = 1$. Thus there are integers r_i and s_i such that $r_im_i + s_in_i = 1$. Let $e_i = s_in_i$. Then $e_i \equiv 1\ (m_i)$ and $e_i \equiv 0\ (m_j)$ for $j \neq i$.

Set $x_0 = \sum_{i=1}^{t} b_ie_i$. Then we have $x_0 \equiv b_ie_i\ (m_i)$ and consequently $x_0 \equiv b_i\ (m_i)$. x_0 is a solution.

Suppose that x_1 is another solution. Then $x_1 - x_0 \equiv 0\ (m_i)$ for $i = 1, 2, \ldots, t$. In other words, $m_1, m_2, \ldots, m_t$ divide $x_1 - x_0$. By Lemma 2, m divides $x_1 - x_0$.

We wish to interpret Theorem 1 from a ring-theoretic point of view. If $R_1, R_2, \ldots, R_n$ are rings, then $R_1 \oplus R_2 \oplus \cdots \oplus R_n = S$, the direct sum of the R_i, is defined to be the set of n-tuples $(r_1, r_2, \ldots, r_n)$ with $r_i \in R_i$. Addition and multiplication are defined by $(r_1, r_2, \ldots, r_n) + (r_1', r_2', \ldots, r_n') = (r_1 + r_1', \ldots, r_n + r_n')$ and $(r_1, r_2, \ldots, r_n) \cdot (r_1', r_2', \ldots, r_n') = (r_1r_1', r_2r_2', \ldots, r_nr_n')$. The zero element is $(0, 0, \ldots, 0)$ and the identity is $(1, 1, \ldots, 1)$. $u \in S$ is a unit iff there is a $v \in S$ such that $uv = 1$.

If $u = (u_1, \ldots, u_n)$ and $v = (v_1, \ldots, v_n)$, then $uv = 1$ implies that $u_i v_i = 1$ for $i = 1, \ldots, n$. Thus u_i is a unit for each i. Conversely, if u_i is a unit for each i, then $u = (u_1, u_2, \ldots, u_n)$ is a unit. For a ring R we denote the group of units by $U(R)$. $U(R_1) \times U(R_2) \times \cdots \times U(R_n)$ is the set of n-tuples $(u_1, u_2, \ldots, u_n)$, where $u_i \in R_i$. This is a group under component-wise multiplication. We have shown

Proposition 3.4.1
If $S = R_1 \oplus R_2 \oplus \cdots \oplus R_n$, then $U(S) = U(R_1) \times U(R_2) \times U(R_3) \times \cdots \times U(R_n)$.

Let $m_1, m_2, \ldots, m_t$ be pairwise relatively prime integers. ψ_i will denote the natural homomorphism from $\mathbb{Z}$ to $\mathbb{Z}/m_i\mathbb{Z}$. We construct a map ψ from $\mathbb{Z}$ to $\mathbb{Z}/m_1\mathbb{Z} \oplus \mathbb{Z}/m_2\mathbb{Z} \oplus \cdots \oplus \mathbb{Z}/m_t\mathbb{Z}$ as follows: $\psi(n) = (\psi_1(n), \psi_2(n), \ldots, \psi_t(n))$ for all $n \in \mathbb{Z}$. It is easy to check that ψ is a ring homomorphism. What are the kernel and image of ψ?

$(\bar{b}_1, \bar{b}_2, \ldots, \bar{b}_t) = \psi(n)$ iff $\psi_i(n) = \bar{b}_i$ for $i = 1, \ldots, t$; i.e., $n \equiv b_i\ (m_i)$ for $i = 1, \ldots, t$. The Chinese remainder theorem assures us that such an n always exists. Thus ψ is onto.

$\psi(n) = 0$ iff $n \equiv 0\ (m_i)$, $i = 1, \ldots, t$, iff n is divisible by $m = m_1 m_2 \cdots m_t$. This is immediate from Lemma 2. Thus the kernel of ψ is the ideal $m\mathbb{Z}$.

We have shown

Theorem 1′
The map ψ induces an isomorphism between $\mathbb{Z}/m\mathbb{Z}$ and $\mathbb{Z}/m_1\mathbb{Z} \oplus \mathbb{Z}/m_2\mathbb{Z} \oplus \cdots \oplus \mathbb{Z}/m_t\mathbb{Z}$.

Corollary
$U(\mathbb{Z}/m\mathbb{Z}) \approx U(\mathbb{Z}/m_1\mathbb{Z}) \times U(\mathbb{Z}/m_2\mathbb{Z}) \times \cdots \times U(\mathbb{Z}/m_t\mathbb{Z})$.

PROOF
Immediate from Theorem 1 and Proposition 3.4.1.

Both sides of the isomorphism in the above corollary are finite groups. The order of the left-hand side is $\phi(m)$ and the order of the right-hand side is $\phi(m_1)\phi(m_2)\cdots\phi(m_t)$. Thus $\phi(m) = \phi(m_1)\phi(m_2)\cdots\phi(m_t)$.

Let $m = p_1^{a_1} p_2^{a_2} \cdots p_t^{a_t}$ be the prime decomposition of m. We have $\phi(m) = \phi(p_1^{a_1})\phi(p_2^{a_2})\cdots\phi(p_t^{a_t})$. For a prime power, p^a, $\phi(p^a) = p^a - p^{a-1}$, because the numbers less than p^a and prime to p^a are prime to p. Since $p^a/p = p^{a-1}$ numbers less than p^a are divisible by p, $p^a - p^{a-1}$ numbers are prime to p. Notice that $p^a - p^{a-1} = p^a(1 - 1/p)$. It follows that $\phi(m) = m \prod (1 - 1/p)$. We proved this formula in Chapter 2 in a much more artificial manner.

Let us summarize. In treating a number of arithmetical questions, the notion of congruence is extremely useful. This notion led us to consider the ring $\mathbb{Z}/m\mathbb{Z}$ and its group of units $U(\mathbb{Z}/m\mathbb{Z})$. To go more deeply into the structure of these algebraic objects we write $m = p_1^{a_1}p_2^{a_2}\cdots p_t^{a_t}$ and are led, via the Chinese remainder theorem, to the following isomorphisms:

$$\mathbb{Z}/m\mathbb{Z} \approx \mathbb{Z}/p_1^{a_1}\mathbb{Z} \oplus \mathbb{Z}/p_2^{a_2}\mathbb{Z} \oplus \cdots \oplus \mathbb{Z}/p_t^{a_t}\mathbb{Z}$$

$$U(\mathbb{Z}/m\mathbb{Z}) \approx U(\mathbb{Z}/p_1^{a_1}\mathbb{Z}) \times U(\mathbb{Z}/p_2^{a_2}\mathbb{Z}) \times \cdots \times U(\mathbb{Z}/p_t^{a_t}\mathbb{Z})$$

For prime powers it is possible to push the investigation much further. This is the subject of Chapter 4.

Notes

It would be useful for the reader to consult other treatments of the basic material given here. See, for example, the very readable book of Davenport [22] and (again) Hardy and Wright [40]. See also Niven and Zuckerman [61], T. Nagell [60], E. Landau [52] and Vinogradov [77].

An interesting discussion of the various possible ways of arranging this material can be found in P. Samuel, Sur l'organization d'un cours d'arithmetique, *L'Enseignment Math.*, **13**, 1967, 223–231. A more advanced discussion of congruences is given in the first chapter of Borevich and Shafarevich [9]; this book also shows how the theory of congruences is useful in determining whether equations can be solved in integers. We mention also the beautiful treatment by J. P. Serre [69].

Historically the notion of congruences was first introduced and used systematically in Gauss's *Disquisitiones Arithmeticae*. The notion of congruence is a wonderful example of the usefulness of employing the "right" notation.

As far as the Chinese remainder theorem is concerned we note that Hardy and Wright [40] note that R. Bachman [4] notes that Sun Tsu was aware of this result in the first century A. D. The theorem is capable of vast generalizations. Properly formulated it holds in any ring with identity. Surprisingly it is no more difficult to prove in general than in the special case we have given.

Exercises

1 Show that there are infinitely many primes congruent to -1 modulo 6.

2 Construct addition and multiplication tables for $\mathbb{Z}/5\mathbb{Z}$, $\mathbb{Z}/8\mathbb{Z}$, and $\mathbb{Z}/10\mathbb{Z}$.

3 Let abc be the decimal representation for an integer between 1 and 1000. Show that abc is divisible by 3 iff $a + b + c$ is divisible by 3. Show that the same result is true if we replace 3 by 9. Show that abc is divisible by 11 iff $a - b + c$ is divisible by 11. Generalize to any number written in decimal notation.

4 Show that the equation $3x^2 + 2 = y^2$ has no solution in integers.

5 Show that the equation $7x^3 + 2 = y^3$ has no solution in integers.

6 Let an integer $n > 0$ be given. A set of integers $a_1, a_2, \ldots, a_{\phi(n)}$ is called a reduced residue system modulo n if they are pairwise incongruent modulo n and $(a_i, n) = 1$ for all i. If $(a, n) = 1$, prove that $aa_1, aa_2, \ldots, aa_{\phi(n)}$ is again a reduced residue system modulo n.

7 Use Exercise 6 to give another proof of Euler's theorem, $a^{\phi(n)} \equiv 1\ (n)$ for $(a, n) = 1$.

8 Let p be an odd prime. If $k \in \{1, 2, \ldots, p-1\}$, show that there is a unique b_k in this set such that $kb_k \equiv 1\ (p)$. Show that $k \neq b_k$ unless $k = 1$ or $k = p - 1$.

9 Use Exercise 7 to prove that $(p-1)! \equiv -1\ (p)$. This is known as Wilson's theorem.

10 If n is not a prime, show that $(n-1)! \equiv 0\ (n)$, except when $n = 4$.

11 Let $a_1, a_2, \ldots, a_{\phi(n)}$ be a reduced residue system modulo n and let N be the number of solutions to $x^2 \equiv 1\ (n)$. Prove that $a_1 a_2 \cdots a_{\phi(n)} \equiv (-1)^{N/2}\ (n)$.

12 Let $\binom{p}{k} = p!/(k!(p-k)!)$ be a binomial coefficient, and suppose that p is a prime. If $1 \leq k \leq p - 1$, show that p divides $\binom{p}{k}$. Deduce $(a+1)^p \equiv a^p + 1\ (p)$.

13 Use Exercise 12 to give another proof of Fermat's theorem, $a^{p-1} \equiv 1\ (p)$ if $p \nmid a$.

14 Let p and q be distinct odd primes such that $p - 1$ divides $q - 1$. If $(n, pq) = 1$, show that $n^{q-1} \equiv 1\ (pq)$.

15 For any prime p show that the numerator of $1 + \frac{1}{2} + \frac{1}{3} + \cdots + 1/p - 1$ is divisible by p. (*Hint:* Make use of Exercises 8 and 9.)

16 Use the proof of the Chinese remainder theorem to solve the system $x \equiv 1\ (7)$, $x \equiv 4\ (9)$, $x \equiv 3\ (5)$.

17 Let $f(x) \in Z[x]$ and $n = p_1^{a_1} p_2^{a_2} \cdots p_t^{a_t}$. Show that $f(x) \equiv 0\ (n)$ has a solution iff $f(x) \equiv 0\ (p_i^{a_i})$ has a solution for $i = 1, 2, \ldots, t$.

18 Let N be the number of solutions to $f(x) \equiv 0\ (n)$ and N_i be the number of solutions to $f(x) \equiv 0\ (p_i^{a_i})$. Prove that $N = N_1 N_2 \cdots N_t$.

19 If p is an odd prime, show that 1 and -1 are the only solutions to $x^2 \equiv 1\ (p^a)$.

20 Show that $x^2 \equiv 1\ (2^b)$ has one solution if $b = 1$, two solutions if $b = 2$, and four solutions if $b \geq 3$.

21 Use Exercises 18–20 to find the number of solutions to $x^2 \equiv 1\ (n)$.

22 Formulate and prove the Chinese remainder theorem in a principal ideal domain.

23 Extend the notion of congruence to the ring $\mathbb{Z}[i]$ and prove that $a + bi$ is always congruent to 0 or 1 modulo $1 + i$.

24 Extend the notion of congruence to the ring $\mathbb{Z}[\omega]$ and prove that $a + b\omega$ is always congruent to either -1, 1, or 0 modulo $1 - \omega$.

25 Let $\lambda = 1 - \omega \in \mathbb{Z}[\omega]$. If $\alpha \in \mathbb{Z}[\omega]$ and $\alpha \equiv 1\ (\lambda)$, prove that $\alpha^3 \equiv 1\ (9)$. (*Hint:* Show first that $3 = -\omega^2\lambda^2$.)

26 Use Exercise 25 to show that if $\xi, \eta, \zeta \in \mathbb{Z}[\omega]$ are not zero and $\xi^3 + \eta^3 + \zeta^3 = 0$, then λ divides at least one of the elements ξ, η, ζ.

chapter four/THE STRUCTURE OF $U(\mathbb{Z}/n\mathbb{Z})$

Having introduced the notion of congruence and discussed some of its properties and applications we shall now go more deeply into the subject. The key result is the existence of primitive roots modulo a prime. This theorem was used by mathematicians before Gauss but he was the first to give a proof. In the terminology introduced in Chapter 3 the existence of primitive roots is equivalent to the fact that $U(\mathbb{Z}/p\mathbb{Z})$ is a cyclic group when p is a prime. Using this fact we shall find an explicit description of the group $U(\mathbb{Z}/n\mathbb{Z})$ for arbitrary n.

I PRIMITIVE ROOTS AND THE GROUP STRUCTURE OF $U(\mathbb{Z}/n\mathbb{Z})$

If $n = p_1^{a_1} p_2^{a_2} \cdots p_l^{a_l}$, then, as was shown in Chapter 3, $U(\mathbb{Z}/n\mathbb{Z}) \approx U(\mathbb{Z}/p_1^{a_1}\mathbb{Z}) \times \cdots \times U(\mathbb{Z}/p_l^{a_l}\mathbb{Z})$. Thus to determine the structure of $U(\mathbb{Z}/n\mathbb{Z})$ it is sufficient to consider the case $U(\mathbb{Z}/p^a\mathbb{Z})$, where p is a prime. We begin by considering the simplest case, $U(\mathbb{Z}/p\mathbb{Z})$.

Since $\mathbb{Z}/p\mathbb{Z}$ is a field, it will be helpful to have available the following simple lemma about fields.

Lemma 1

Let $f(x) \in k[x]$, k a field. Suppose that $\deg f(x) = n$. Then f has at most n distinct roots.

PROOF

The proof goes by induction on n. For $n = 1$ the assertion is trivial. Assume that the lemma is true for polynomials of degree $n - 1$. If $f(x)$ has no roots in k, we are done. If α is a root, $f(x) = q(x)(x - \alpha) + r$, where r is a constant. Setting $x = \alpha$ we see that $r = 0$. Thus $f(x) = q(x)(x - \alpha)$ and $\deg q(x) = n - 1$. If $\beta \neq \alpha$ is another root of $f(x)$, then $0 = f(\beta) = (\beta - \alpha)q(\beta)$, which implies that $q(\beta) = 0$. Since by induction $q(x)$ has at most $n - 1$ distinct roots, $f(x)$ has at most n distinct roots.

Corollary

Let $f(x), g(x) \in k[x]$ and $\deg f(x) = \deg g(x) = n$. If $f(\alpha_i) = g(\alpha_i)$ for $n + 1$ distinct elements $\alpha_1, \alpha_2, \ldots, \alpha_n, \alpha_{n+1}$, then $f(x) = g(x)$.

PROOF

Apply the lemma to the polynomial $f(x) - g(x)$.

Proposition 4.1.1

$x^{p-1} - 1 \equiv (x-1)(x-2)\cdots(x-p+1)\ (p)$.

PROOF

If $\bar{a}$ denotes the residue class of an integer a in $\mathbb{Z}/p\mathbb{Z}$, an equivalent way of stating the proposition is $x^{p-1} - \bar{1} = (x - \bar{1})(x - \bar{2})\cdots(x - (\overline{p-1}))$ in $\mathbb{Z}/p\mathbb{Z}[x]$. Let $f(x) = (x^{p-1} - \bar{1}) - (x - \bar{1})(x - \bar{2})\cdots(x - (\overline{p-1}))$. $f(x)$ has degree less than $p - 1$ (the leading terms cancel) and has the $p - 1$ roots $\bar{1}, \bar{2}, \ldots, \overline{p-1}$ (Fermat's little theorem). Thus $f(x)$ is identically zero.

Corollary

$(p-1)! \equiv -1\ (p)$.

PROOF

Set $x = 0$ in Proposition 4.1.1.

This result is known as Wilson's theorem. It is not hard to prove that if $n > 4$ is not prime, then $(n-1)! \equiv 0\ (n)$ (see Exercise 10 of Chapter 3). Thus the congruence $(n-1)! \equiv -1\ (n)$ is characteristic for primes. We shall make use of Wilson's theorem later when discussing quadratic residues.

Proposition 4.1.2

If $d \mid p - 1$, then $x^d \equiv 1\ (p)$ has exactly d solutions.

PROOF

Let $dd' = p - 1$. Then

$$\frac{x^{p-1} - 1}{x^d - 1} = \frac{(x^d)^{d'} - 1}{x^d - 1} = (x^d)^{d'-1} + (x^d)^{d'-2} + \cdots + x^d + 1 = g(x)$$

Therefore

$$x^{p-1} - 1 = (x^d - 1)g(x)$$

and

$$x^{p-1} - \bar{1} = (x^d - \bar{1})\bar{g}(x)$$

If $x^d - \bar{1}$ had less than d roots, then by Lemma 1 the right-hand side would have less than $p - 1$ roots. However, the left-hand side has the

$p - 1$ roots $\bar{1}, \bar{2}, \ldots, \overline{p-1}$. Thus $x^d \equiv 1\ (p)$ has exactly d roots as asserted.

Theorem 1
$U(\mathbb{Z}/p\mathbb{Z})$ is a cyclic group.

PROOF
For $d \mid p - 1$ let $\psi(d)$ be the number of elements in $U(\mathbb{Z}/p\mathbb{Z})$ of order d. By Proposition 4.1.2 we see that the elements of $U(\mathbb{Z}/p\mathbb{Z})$ satisfying $x^d \equiv \bar{1}$ form a group of order d. Thus $\sum_{c|d} \psi(c) = d$. Applying the Möbius inversion theorem we obtain $\psi(d) = \sum_{c|d} \mu(c)d/c$. The right-hand side of this equation is equal to $\phi(d)$, as was seen in the proof of Proposition 2.2.5. In particular, $\psi(p - 1) = \phi(p - 1)$, which is greater than 1 if $p > 2$. Since the case $p = 2$ is trivial, we have shown in all cases the existence of an element [in fact, $\phi(p - 1)$ elements] of order $p - 1$.

Theorem 1 is of fundamental importance. It was first proved by Gauss. After giving some new terminology we shall outline two more proofs.

Definition
An integer a is called a *primitive root* mod p if $\bar{a}$ generates the group $U(\mathbb{Z}/p\mathbb{Z})$. Equivalently, a is a primitive root mod p if $p - 1$ is the smallest positive integer such that $a^{p-1} \equiv 1\ (p)$.

As an example, 2 is a primitive root mod 5, since the least positive residues of 2, 2^2, 2^3, and 2^4 are 2, 4, 3, and 1. Thus $4 = 5 - 1$ is the smallest positive integer such that $2^n \equiv 1\ (5)$.

For $p = 7$, 2 is not a primitive root since $2^3 \equiv 1\ (7)$, but 3 is since $3, 3^2, 3^3, 3^4, 3^5$, and 3^6 are congruent to 3, 2, 6, 4, 5, and 1 mod 7.

Although Theorem 1 shows the existence of primitive roots for a given prime, there is no simple way of finding one. For small primes trial and error is probably as good a method as any.

A celebrated conjecture of E. Artin states that if $a > 1$ is not a square, then there are infinitely many primes for which a is a primitive root. Some progress has been made in recent years, but the conjecture still seems far from resolution. See [35].

Because of its importance, we outline two more proofs of Theorem 1. The reader is invited to fill in the details.

Let $p - 1 = q_1^{e_1} q_2^{e_2} \cdots q_t^{e_t}$ be the prime decomposition of $p - 1$. Consider the congruences

1. $x^{q_i^{e_i - 1}} \equiv 1\ (p)$.
2. $x^{q_i^{e_i}} \equiv 1\ (p)$.

Every solution to congruence 1 is a solution of congruence 2. Moreover, congruence 2 has more solutions than congruence 1. Let g_i be a solution to congruence 2 that is not a solution to congruence 1 and set $g = g_1 g_2 \cdots g_t$. $\bar{g}_i$ generates a subgroup of $U(\mathbb{Z}/p\mathbb{Z})$ of order $q_i^{e_i}$. It follows that $\bar{g}$ generates a subgroup of $U(\mathbb{Z}/p\mathbb{Z})$ of order $q_1^{e_1} q_2^{e_2} \cdots q_t^{e_t} = p - 1$. Thus g is a primitive root and $U(\mathbb{Z}/p\mathbb{Z})$ is cyclic.

Finally, on group-theoretic grounds we can see that $\psi(d) \leq \phi(d)$ for $d \mid p - 1$. Both $\sum_{d|p-1} \psi(d)$ and $\sum_{d|p-1} \phi(d)$ are equal to $p - 1$. It follows that $\psi(d) = \phi(d)$ for all $d \mid p - 1$. In particular, $\psi(p - 1) = \phi(p - 1)$. For $p > 2$, $\phi(p - 1) > 1$, implying that $\psi(p - 1) > 1$. The result follows.

The notion of primitive root can be generalized somewhat.

Definition

Let $a, n \in \mathbb{Z}$. a is said to be a *primitive root* mod n if the residue class of a mod n generates $U(\mathbb{Z}/n\mathbb{Z})$. It is equivalent to require that a and n be relatively prime and that $\phi(n)$ be the smallest positive integer such that $a^{\phi(n)} \equiv 1\ (n)$.

In general, it is not true that $U(\mathbb{Z}/n\mathbb{Z})$ is cyclic. For example, the elements of $U(\mathbb{Z}/8\mathbb{Z})$ are $\bar{1}, \bar{3}, \bar{5}, \bar{7}$, and $\bar{1}^2 = \bar{1}$, $\bar{3}^2 = \bar{1}$, $\bar{5}^2 = \bar{1}$, and $\bar{7}^2 = \bar{1}$. Thus there is no element of order $4 = \phi(8)$. It follows that not every integer possesses primitive roots. We shall shortly determine those integers that do.

Lemma 2

If p is a prime and $1 \leq k < p$, then the binomial coefficient $\binom{p}{k}$ is divisible by p.

PROOF

We give two proofs.

(a) By definition $\binom{p}{k} = \dfrac{p!}{k!(p-k)!}$ so that $p! = k!(p-k)!\binom{p}{k}$. Now, p divides $p!$, but p does not divide $k!(p - k)!$ since this expression is a product of integers less than, and thus relatively prime to, p. Thus p divides $\binom{p}{k}$.

(b) By Fermat's little theorem $a^{p-1} \equiv 1\ (p)$ if $p \nmid a$. It follows that $a^p \equiv a\ (p)$ for all a. In particular, $(1 + a)^p \equiv 1 + a \equiv 1 + a^p\ (p)$ for all a. Thus $(1 + x)^p - 1 - x^p \equiv 0\ (p)$ has p solutions. Since the polynomial has degree less than p it follows from the corollary to Lemma 1 that $(\bar{1} + x)^p - \bar{1} - x^p$ is identically zero in $\mathbb{Z}/p\mathbb{Z}[x]$. $(1 + x)^p - 1 - x^p = \sum_{k=1}^{p-1} \binom{p}{k} x^k$. Thus $\binom{p}{k} = \bar{0}$ for $1 \leq k \leq p - 1$, implying that $p \mid \binom{p}{k}$. The only interest in this proof is that we do not assume any information on $\binom{p}{k}$.

Lemma 3

If $l \geq 1$ and $a \equiv b\ (p^l)$, then $a^p \equiv b^p\ (p^{l+1})$.

PROOF

We may write $a = b + cp^l, c \in \mathbb{Z}$. Thus $a^p = b^p + \binom{p}{1} b^{p-1}cp^l + A$, where A is an integer divisible by p^{l+2}. The second term is clearly divisible by p^{l+1}. Thus

$$a^p \equiv b^p\ (p^{l+1})$$

Corollary 1

If $l \geq 2$ and $p \neq 2$, then $(1 + ap)^{p^{l-2}} \equiv 1 + ap^{l-1}(p^l)$ for all $a \in \mathbb{Z}$.

PROOF

The proof is by induction on l. For $l = 2$ the assertion is trivial. Suppose that it is true for some $l \geq 2$. We show that it is then true for $l + 1$. Applying Lemma 3 we obtain

$$(1 + ap)^{p^{l-1}} \equiv (1 + ap^{l-1})^p (p^{l+1})$$

By the binomial theorem

$$(1 + ap^{l-1})^p = 1 + \binom{p}{1} ap^{l-1} + B$$

where B is a sum of $p - 2$ terms. Using Lemma 2 it is easy to see that all these terms are divisible by $p^{1+2(l-1)}$ except perhaps for the last term, $a^p p^{p(l-1)}$. Since $l \geq 2, 1 + 2(l - 1) \geq l + 1$, and since also $p \geq 3$, $p(l - 1) \geq l + 1$. Thus $p^{l+1} \mid B$ and $(1 + ap)^{p^{l-1}} \equiv 1 + ap^l$ (p^{l+1}), which is as required.

Before starting a second corollary we need a definition.

Definition

Let $a, n \in \mathbb{Z}$ and $(a, n) = 1$. We say a has *order* e mod n if e is the smallest positive integer such that $a^e \equiv 1\ (n)$. This is equivalent to saying that $\bar{a}$ has order e in the group $U(\mathbb{Z}/n\mathbb{Z})$.

Corollary 2
If $p \neq 2$ and $p \nmid a$, then p^{l-1} is the order of $1 + ap \bmod p^l$.

PROOF
By Corollary 1, $(1 + ap)^{p^{l-1}} \equiv 1 + ap^l \ (p^{l+1})$, implying that $(1 + ap)^{p^{l-1}} \equiv 1 \ (p^l)$ and thus that $1 + ap$ has order dividing p^{l-1}. $(1 + ap)^{p^{l-2}} \equiv 1 + ap^{l-1} \ (p^l)$ shows that p^{l-2} is not the order of $1 + ap$ (it is here we use the hypothesis $p \nmid a$). The result follows.

We are now in a position to extend Theorem 1. It turns out that we shall have to treat the prime 2 separately from the odd primes. The necessity of treating 2 differently from the other primes occurs repeatedly in number theory.

Theorem 2
If p is an odd prime and $l \in \mathbb{Z}^+$, then $U(\mathbb{Z}/p^l\mathbb{Z})$ is cyclic; i.e., there exist primitive roots mod p^l.

PROOF
By Theorem 1 there exist primitive roots mod p. If $g \in Z$ is a primitive root mod p, then so is $g + p$. If $g^{p-1} \equiv 1 \ (p^2)$, then $(g + p)^{p-1} \equiv g^{p-1} + (p-1)g^{p-2}p \equiv 1 + (p-1)g^{p-2}p \ (p^2)$. Since p^2 does not divide $(p-1)g^{p-2}p$ we may assume from the beginning that g is a primitive root mod p and that $g^{p-1} \not\equiv 1 \ (p^2)$.

We claim that such a g is already a primitive root mod p^l. To prove this it is sufficient to prove that if $g^n \equiv 1 \ (p^l)$, then $\phi(p^l) = p^{l-1}(p-1) \mid n$.

$g^{p-1} = 1 + ap$, where $p \nmid a$. By Corollary 2 to Lemma 3, p^{l-1} is the order of $1 + ap \bmod p^l$. Since $(1 + ap)^n \equiv 1 \ (p^l)$ we have $p^{l-1} \mid n$.

Let $n = p^{l-1}n'$. Then $g^n = (g^{p^{l-1}})^{n'} \equiv g^{n'} \ (p)$, and therefore $g^{n'} \equiv 1 \ (p)$. Since g is a primitive root mod p, $p - 1 \mid n'$. We have proved that $p^{l-1}(p-1) \mid n$, as required.

Theorem 2'
2^l has primitive roots for $l = 1$ or 2 but not for $l \geq 3$. If $l \geq 3$, then $\{(-1)^a 5^b \mid a = 0, 1$ and $0 \leq b < 2^{l-2}\}$ constitutes a reduced residue system mod 2^l. *It follows that for $l \geq 3$, $U(\mathbb{Z}/2^l\mathbb{Z})$ is the direct product of two cyclic groups, one of order 2, the other of order 2^{l-2}.*

PROOF
1 is a primitive root mod 2, and 3 is a primitive root mod 4. From now on let us assume that $l \geq 3$.

We claim that (1) $5^{2^{l-3}} \equiv 1 + 2^{l-1} \ (2^l)$. This is true for $l = 3$. Assume that it is true for $l \geq 3$ and we shall prove it is true for $l + 1$. First notice that $(1 + 2^{l-1})^2 = 1 + 2^l + 2^{2l-2}$ and that $2l - 2 \geq l + 1$ for

$l \geq 3$. Applying Lemma 3 to congruence (1), we get (2) $5^{2^{l-2}} \equiv 1 + 2^l$ (2^{l+1}). Our claim is now established by induction.

From (2) we see that $5^{2^{l-2}} \equiv 1$ (2^l), whereas from (1) we see that $5^{2^{l-3}} \not\equiv 1$ (2^l). Thus 2^{l-2} is the order of 5 mod 2^l.

Consider the set $\{(-1)^a 5^b \mid a = 1, 2 \text{ and } 0 \leq b < 2^{l-2}\}$. We claim that these 2^{l-1} numbers are incongruent mod 2^l. Since $\phi(2^l) = 2^{l-1}$ this will show that our set is in fact a reduced residue system mod 2^l.

If $(-1)^a 5^b \equiv (-1)^{a'} 5^{b'}$ (2^l), $l \geq 3$, then $(-1)^a \equiv (-1)^{a'}$ (4), implying that $a \equiv a'$ (2). Thus $a = a'$. Going further, $a = a'$ implies that $5^b \equiv 5^{b'}$ (2^l) or that $5^{b-b'} \equiv 1$ (2^l). Therefore, $b \equiv b'$ (2^{l-2}), which yields $b = b'$.

Finally, notice that $(-1)^a 5^b$ raised to the 2^{l-2} power is congruent to 1 mod 2^l. Thus 2^l has no primitive roots if $l \geq 3$.

Consider the situation mod 8. 1, 3, 5, and 7 constitute a reduced residue system. We have $5^0 \equiv 1$, $5^1 \equiv 5$, $-5^0 \equiv 7$, and $-5^1 \equiv 3$. Table 1 represents the situation mod 16. The second row contains the least positive residues of the powers of 5, and the third row those of the negative powers of 5.

Table 1

	5^0	5^1	5^2	5^3
+	1	5	9	13
−	15	11	7	3

Theorems 2 and 2′ permit us to give a fairly complete description of the group $U(\mathbb{Z}/n\mathbb{Z})$ for arbitrary n.

Theorem 3

Let $n = 2^a p_1^{a_1} p_2^{a_2} \cdots p_l^{a_l}$ be the prime decomposition of n. Then

$$U(\mathbb{Z}/n\mathbb{Z}) \approx U(\mathbb{Z}/2^a\mathbb{Z}) \times U(\mathbb{Z}/p_1^{a_1}\mathbb{Z}) \times \cdots \times U(\mathbb{Z}/p_l^{a_l}\mathbb{Z})$$

$U(\mathbb{Z}/p_i^{a_i}\mathbb{Z})$ is a cyclic group of order $p_i^{a_i-1}(p_i - 1)$. $U(\mathbb{Z}/2^a\mathbb{Z})$ is cyclic of order 1 and 2 for $a = 1$ and 2, respectively. If $a \geq 3$, then it is the product of two cyclic groups, one of order 2, the other of order 2^{a-2}.

PROOF

Theorems 2, 2′, and Theorem 1′ of Chapter 3.

We conclude this section by giving an answer to the question of which integers possess primitive roots.

Proposition 4.1.3
n possesses primitive roots iff n is of the form $2, 4, p^a$, or $2p^a$, where p is an odd prime.

PROOF
By Theorem 2′ we can assume that $n \neq 2^l, l \geq 3$. If n is not of the given form, it is easy to see that n can be written as a product $m_1 m_2$, where $(m_1, m_2) = 1$ and $m_1, m_2 > 2$. We then have that $\phi(m_1)$ and $\phi(m_2)$ are both even and that $U(\mathbb{Z}/n\mathbb{Z}) \approx U(\mathbb{Z}/m_1\mathbb{Z}) \times U(\mathbb{Z}/m_2\mathbb{Z})$. Both $U(\mathbb{Z}/m_1\mathbb{Z})$ and $U(\mathbb{Z}/m_2\mathbb{Z})$ have elements of order 2, but this shows that $U(\mathbb{Z}/n\mathbb{Z})$ is not cyclic since a cyclic group contains at most one element of order 2. Thus n does not possess primitive roots.

We already know that 2, 4, and p^a possess primitive roots. Since $U(\mathbb{Z}/2p^a\mathbb{Z}) \approx U(\mathbb{Z}/2\mathbb{Z}) \times U(\mathbb{Z}/p^a\mathbb{Z}) \approx U(\mathbb{Z}/p^a\mathbb{Z})$ it follows that $U(\mathbb{Z}/2p^a\mathbb{Z})$ is cyclic; i.e., $2p^a$ possesses primitive roots.

2 *n*th *POWER RESIDUES*

Definition
If $m, n \in \mathbb{Z}^+$, $a \in \mathbb{Z}$, and $(a, m) = 1$, then we say that a is an nth *power residue* mod m if $x^n \equiv a\ (m)$ is solvable.

Proposition 4.2.1
If $m \in \mathbb{Z}^+$ possesses primitive roots and $(a, m) = 1$, then a is an nth power residue mod m *iff $a^{\phi(m)/d} \equiv 1\ (m)$, where $d = (n, \phi(m))$.*

PROOF
Let g be a primitive root mod m and $a = g^b$, $x = g^y$. Then the congruence $x^n \equiv a\ (m)$ is equivalent to $g^{ny} \equiv g^b\ (m)$, which in turn is equivalent to $ny \equiv b\ (\phi(m))$. The latter congruence is solvable iff $d \mid b$. Moreover, it is useful to notice that if there is one solution, there are exactly d solutions.

If $d \mid b$, then $a^{\phi(m)/d} \equiv g^{b\phi(m)/d} \equiv 1\ (m)$. Conversely, if $a^{\phi(m)/d} \equiv 1\ (m)$, then $g^{b\phi(m)/d} \equiv 1\ (m)$, which implies that $\phi(m)$ divides $b\phi(m)/d$ or $d \mid b$. This proves the result.

The proof yields the following additional information. If $x^n \equiv a\ (m)$ is solvable, there are exactly $(n, \phi(m))$ solutions.

Now suppose that $m = 2^e p_1^{e_1} \cdots p_l^{e_l}$. Then $x^n \equiv a\ (m)$ is solvable iff the system of congruences

$$x^n \equiv a\ (2^e), x^n \equiv a\ (p_1^{e_1}), \ldots, x^n \equiv a\ (p_l^{e_l})$$

is solvable. Since odd prime powers possess primitive roots we may apply Proposition 4.2.1 to the last l congruences. We are reduced to a consideration of the congruence $x^n \equiv a\ (2^e)$. Since 2 and 4 possess primitive roots we may further assume that $e \geq 3$.

Proposition 4.2.2

Suppose that a is odd, $e \geq 3$, and consider the congruence $x^n \equiv a\ (2^e)$. If n is odd, a solution always exists and it is unique.

If n is even, a solution exists iff $a \equiv 1\ (4)$, $a^{2^{e-2}/d} \equiv 1\ (2^e)$, where $d = (n, 2^{e-2})$, When a solution exists there are exactly $2d$ solutions.

PROOF

We leave the proof as an exercise. One begins by writing $a \equiv (-1)^s 5^t (2^e)$ and $x \equiv (-1)^y 5^z\ (2^e)$.

Propositions 4.2.1 and 4.2.2 give a fairly satisfactory answer to the question; When is an integer a an nth power residue mod m? It is possible to go a bit further in some cases.

Proposition 4.2.3

If p is an odd prime, $p \nmid a$, and $p \nmid n$, then if $x^n \equiv a\ (p)$ is solvable, so is $x^n \equiv a\ (p^e)$ for all $e \geq 1$. All these congruences have the same number of solutions.

PROOF

If $n = 1$, the assertion is trivial, so we may assume $n \geq 2$. Suppose that $x^n \equiv a\ (p^e)$ is solvable. Let x_0 be a solution and set $x_1 = x_0 + bp^e$. A short computation shows $x_1^n \equiv x_0^n + nbp^e x_0^{n-1}\ (p^{e+1})$. We wish to solve $x_1^n \equiv a\ (p^{e+1})$. This is equivalent to finding an integer b such that $nx_0^{n-1}b \equiv ((a - x_0^n)/p^e)\ (p)$. Notice that $(a - x_0^n)/p^e$ is an integer and that $p \nmid nx_0^{n-1}$. Thus this congruence is uniquely solvable for b, and with this value of b, $x_1^n \equiv a\ (p^{e+1})$.

If $x^n \equiv a\ (p)$ has no solutions, then $x^n \equiv a\ (p^e)$ has no solutions. On the other hand, if $x^n \equiv a\ (p)$ has a solution, so do all the congruences $x^n \equiv a\ (p^e)$, as we have just seen. By the remark following Proposition 4.2.1 the number of solutions to $x^n \equiv a\ (p^e)$ is $(n, \phi(p^e))$ provided one solution exists. If $p \nmid n$, it is easy to see that $(n, \phi(p)) = (n, \phi(p^e))$ for all $e \geq 1$. This concludes the proof.

As usual the result for the powers of 2 is more complicated.

Proposition 4.2.4

Let 2^l be the highest power of 2 dividing n. Suppose that a is odd and that

$x^n \equiv a\ (2^{2l+1})$ is solvable. Then $x^n \equiv a\ (2^e)$ is solvable for all $e \geq 2l + 1$ (and consequently for all $e \geq 1$). Moreover, all these congruences have the same number of solutions.

PROOF

We leave the proof as an exercise. One begins by assuming that $x^n \equiv a\ (2^m)$, $m \geq 2l + 1$, has a solution x_0. Let $x_1 = x_0 + b2^{m-l}$. One shows, by an appropriate choice of b, that $x_1^n \equiv a\ (2^{m+1})$.

Notice that $x^2 \equiv 5\ (2^2)$ is solvable (for example, $x = 1$) but that $x^2 \equiv 5\ (2^3)$ is not. On the other hand, one can prove easily from the proposition that if $a \equiv 1\ (8)$, then $x^2 \equiv a\ (2^e)$ is solvable for all e and conversely.

Notes

Lemma 1 and its important consequence, Proposition 4.1.1, are due to J. Lagrange (1768).

Fermat's theorem [that $a^{p-1} \equiv 1\ (p)$ if $p \nmid a$] was first proved by Euler. Wilson's theorem was stated by E. Waring and proved by Lagrange.

The important result on the existence of primitive roots modulo a prime was asserted by Euler and, as we have mentioned, was first proved by Gauss. The proofs of this result can be modified to prove the more general assertion that a finite subgroup of the multiplicative group of a field is cyclic, i.e., is generated by one element. This result for the case of the complex numbers is due to A. De Moivre.

There are a number of interesting conjectures related to primitive roots. The celebrated conjecture of E. Artin asserts that given an integer a that is not a square, and not -1, there are infinitely many primes for which a is a primitive root. In the case $a = 10$ this goes back to Gauss and amounts to requiring the existence of infinitely many primes p such that the period of the decimal expansion of $1/p$ has length $p - 1$. (See Chapter 4 of Rademacher [64] for a quick introduction to the theory of decimal expansions.) An excellent survey article devoted to the Artin conjecture and related questions has recently appeared (Goldstein [35]).

Lehmer [54] discovered the following curious result. The first prime of the form $326n^2 + 3$ for which 326 is not a primitive root must be bigger than 10 million. He mentions other results of the same nature. It would be interesting to see what is responsible for this strange behavior.

Given a prime p, what can be said about the size of the smallest positive integer that is a primitive root mod p? This problem has given rise to a lot of research. One contribution, due to L. K. Hua, is that the number in question is less than $2^{m+1}p^{1/2}$, where m is the number of distinct primes dividing $p - 1$. For a discussion of this problem and a good bibliography, see Erdös [31]. For other interesting results and problems see [76] and [12].

There exist many investigations into the existence of sequences of consecutive integers each of which is a kth power modulo p. Consider primes of the form $kt + 1$. A basic result due to A. Brauer asserts that if m is a given positive integer, then for all primes p sufficiently large there are m consecutive integers $r, r + 1, \ldots, r + m - 1$ all of which are kth powers modulo p. The question of finding the least such r for given p and m is a problem of current interest. For this, and a discussion of other open questions in this area, see the article by Mills [59].

Given a prime p, what can be said about the size of the smallest positive integer that is a nonsquare modulo p? An interesting conjecture is the following: For a given n the integer in question is smaller than $\sqrt[n]{p}$ for all sufficiently large p. For more discussion, see P. Erdös [31] and Chapter 3 of Chowla [18].

Finally, we mention that an analog of the Artin conjecture on primitive roots has actually been proved in the ring $k[x]$ by H. Bilharz [8]. Bilharz proved his theorem under the assumption that the Riemann hypothesis held for the so-called congruence zeta function (see Chapter 11). This was actually proved several years later by A. Weil. In recent years C. Hooley was able to prove that Artin's original conjecture was correct under the assumption that the extended Riemann hypothesis held in algebraic number fields [46]. For a discussion of the classical Riemann hypothesis and its consequences, see Chowla [18]. No one at present seems to have the slightest idea as to how to prove the Riemann hypothesis for number fields so that it seems clear that Hooley is not about to have the same good luck that Bilharz enjoyed.

Exercises

1 Show that 2 is a primitive root modulo 29.

2 Compute all primitive roots for $p = 11, 13, 17$, and 19.

3 Suppose that a is a primitive root modulo p^n, p an odd prime. Show that a is a primitive root modulo p.

4 Consider a prime p of the form $4t + 1$. Show that a is a primitive root modulo p iff $-a$ is a primitive root modulo p.

5 Consider a prime p of the form $4t + 3$. Show that a is a primitive root modulo p iff $-a$ has order $(p - 1)/2$.

6 If $p = 2^n + 1$ is a Fermat prime, show that 3 is a primitive root modulo p.

7 Suppose that p is a prime of the form $8t + 3$ and that $q = (p - 1)/2$ is also a prime. Show that 2 is a primitive root modulo p.

8 Let p be an odd prime. Show that a is a primitive root modulo p iff $a^{(p-1)/q} \not\equiv 1$ (p) for all prime divisors q of $p-1$.
9 Show that the product of all the primitive roots modulo p is congruent to $(-1)^{\phi(p-1)}$ modulo p.
10 Show that the sum of all the primitive roots modulo p is congruent to $\mu(p-1)$ modulo p.
11 Prove that $1^k + 2^k + \cdots + (p-1)^k \equiv 0$ (p) if $p - 1 \nmid k$ and -1 (p) if $p - 1 \mid k$.
12 Use the existence of a primitive root to give another proof of Wilson's theorem $(p-1)! \equiv -1$ (p).
13 Let G be a finite cyclic group and $g \in G$ a generator. Show that all the other generators are of the form g^k, where $(k, n) = 1$, n being the order of G.
14 Let A be a finite abelian group and $a, b \in A$ elements of order m and n, respectively. If $(m, n) = 1$, prove that ab has order mn.
15 Let K be a field and $G \subseteq K^*$ a finite subgroup of the multiplicative group of K. Extend the arguments used in the proof of Theorem 1 to show that G is cyclic.
16 Calculate the solutions to $x^3 \equiv 1$ (19) and $x^4 \equiv 1$ (17).
17 Use the fact that 2 is a primitive root modulo 29 to find the seven solutions to $x^7 \equiv 1$ (29).
18 Solve the congruence $1 + x + x^2 + \cdots + x^6 \equiv 0$ (29).
19 Determine the numbers a such that $x^3 \equiv a$ (p) is solvable for $p = 7, 11$, and 13.
20 Let p be a prime and d a divisor of $p-1$. Show that the dth powers form a subgroup of $U(\mathbb{Z}/p\mathbb{Z})$ of order $(p-1)/d$. Calculate this subgroup for $p = 11, d = 5$; $p = 17, d = 4$; $p = 19, d = 6$.
21 If g is a primitive root modulo p and $d \mid p - 1$, show that $g^{(p-1)/d}$ has order d. Show also that a is a dth power iff $a \equiv g^{kd}$ (p) for some k. Do Exercises 16–20 making use of these observations.
22 If a has order 3 modulo p, show that $1 + a$ has order 6.
23 Show that $x^2 \equiv -1$ (p) has a solution iff $p \equiv 1$ (4) and that $x^4 \equiv -1$ (p) has a solution iff $p \equiv 1$ (8).
24 Show that $ax^m + by^n \equiv c$ (p) has the same number of solutions as $ax^{m'} + by^{n'} \equiv c$ (p), where $m' = (m, p-1)$ and $n' = (n, p-1)$.
25 Prove Propositions 4.2.2 and 4.2.4.

chapter five/QUADRATIC RECIPROCITY

If p is a prime, the discussion of the congruence $x^2 \equiv a\,(p)$ is fairly easy. It is solvable iff $a^{(p-1)/2} \equiv 1\ (p)$. With this fact in hand a complete analysis is a simple matter. However, if the question is turned around, the problem is much more difficult. Suppose that a is an integer. For which primes p is the congruence $x^2 \equiv a\ (p)$ solvable? The answer is provided by the law of quadratic reciprocity. This law was formulated by Euler and A. M. Legendre but Gauss was the first to provide a complete proof. Gauss was extremely proud of this result. He called it the Theorema Aureum, *the golden theorem.*

I *QUADRATIC RESIDUES*

If $(a, m) = 1$, a is called a quadratic residue mod m if the congruence $x^2 \equiv a\,(m)$ has a solution. Otherwise a is called a quadratic nonresidue mod m.

For example, 2 is a quadratic residue mod 7, but 3 is not. In fact, $1^2, 2^2, 3^2, 4^2, 5^2$, and 6^2 are congruent to 1, 4, 2, 2, 4, and 1, respectively. Thus 1, 2, and 4 are quadratic residues, and 3, 5, and 6 are not.

Given any fixed positive integer m it is possible to determine the quadratic residues by simply listing the positive integers less than and prime to m, squaring them, and reducing mod m. This is what we have just done for $m = 7$.

The following proposition gives a less tedious way of deciding when a given integer is a quadratic residue mod m.

Proposition 5.1.1

Let $m = 2^e p_1^{e_1} \cdots p_l^{e_l}$ be the prime decomposition of m, and suppose that $(a, m) = 1$. Then $x^2 \equiv a\ (m)$ is solvable iff the following conditions are satisfied:

(a) *If $e = 2$, then $a \equiv 1\ (4)$.*
   If $e \geq 3$, then $a \equiv 1\ (8)$.

(b) *For each i we have $a^{(p_i-1)/2} \equiv 1\ (p_i)$.*

PROOF

By the Chinese remainder theorem the congruence $x^2 \equiv a\ (m)$ is equivalent to the system $x^2 \equiv a\,(2^e), x^2 \equiv a\,(p_1^{e_1}), \ldots, x^2 \equiv a\,(p_l^{a_l})$.

Consider $x^2 \equiv a\ (2^e)$. 1 is the only quadratic residue mod 4, and 1 is the only quadratic residue mod 8. Thus we have solvability iff $a \equiv 1\ (4)$ if $e = 2$ and $a \equiv 1\ (8)$ if $e = 3$. A direct application of Proposition 4.2.4 shows that $x^2 \equiv a\ (8)$ is solvable iff $x^2 \equiv a\ (2^e)$ is solvable for all $e \geq 3$.

Now consider $x^2 \equiv a\ (p_i^{e_i})$. Since $(2, p_i) = 1$ it follows from Proposition 4.2.3 that this congruence is solvable iff $x^2 \equiv a\ (p_i)$ is solvable. To this congruence apply Proposition 4.2.1 with $n = 2$, $m = p$, and $d = (n, \phi(m)) = (2, p - 1) = 2$. We obtain that $x^2 \equiv a\ (p_i)$ is solvable iff $a^{(p_i - 1)/2} \equiv 1\ (p_i)$.

This result reduces questions about quadratic residues to the corresponding questions for prime moduli. In what follows p will denote an odd prime.

Definition

The symbol (a/p) will have the value 1 if a is a quadratic residue mod p, -1 if a is a quadratic nonresidue mod p, and zero if $p \mid a$. (a/p) is called the *Legendre symbol.*

The Legendre symbol is an extremely convenient device for discussing quadratic residues. We shall list some of its properties.

Proposition 5.1.2

(a) $a^{(p-1)/2} \equiv (a/p)\ (p)$.
(b) $(ab/p) = (a/p)(b/p)$.
(c) *If* $a \equiv b\ (p)$, *then* $(a/p) = (b/p)$.

PROOF

If p divides a or b, all three assertions are trivial. Assume that $p \nmid a$ and that $p \nmid b$.

We know that $a^{p-1} \equiv 1\ (p)$; thus $(a^{(p-1)/2} + 1)(a^{(p-1)/2} - 1) = a^{p-1} - 1 \equiv 0\ (p)$. It follows that $a^{(p-1)/2} \equiv \pm 1\ (p)$. By Proposition 5.1.1, $a^{(p-1)/2} \equiv 1\ (p)$ iff a is a quadratic residue mod p. This proves part (a).

To prove part (b) we apply part (a). $(ab)^{(p-1)/2} \equiv (ab/p)\ (p)$ and $(ab)^{(p-1)/2} = a^{(p-1)/2}b^{(p-1)/2} \equiv (a/p)(b/p)\ (p)$. Thus $(ab/p) \equiv (a/p)(b/p)$ (p), which implies that $(ab/p) = (a/p)(b/p)$.

Part (c) is obvious from the definition.

Corollary 1

There are as many residues as nonresidues mod p.

PROOF

$a^{(p-1)/2} \equiv 1\ (p)$ has $(p - 1)/2$ solutions. Thus there are $(p - 1)/2$ residues and $p - 1 - ((p - 1)/2) = (p - 1)/2$ nonresidues.

Corollary 2

The product of two residues is a residue, the product of two nonresidues is a residue, and the product of a residue and a nonresidue is a nonresidue.

PROOF

This all follows easily from part (b).

Corollary 3

$(-1)^{(p-1)/2} = (-1/p)$.

PROOF

Substitute $a = -1$ in part (a).

Corollary 3 is particularly interesting. Every odd integer has the form $4k + 1$ or $4k + 3$. Using this one can restate Corollary 3 as follows: $x^2 \equiv -1\ (p)$ has a solution iff p is of the form $4k + 1$. Thus -1 is a residue of the primes $5, 13, 17, 29, \ldots$ and a nonresidue of the primes $3, 7, 11, 19, \ldots$. The reader should check some of these assertions numerically.

One is led by this result to ask a more general question. If a is an integer, for which primes p is a quadratic residue mod p? The answer to this question is provided by the law of quadratic reciprocity to whose statement and proof we shall soon devote a great deal of attention.

Corollary 3 enables us to prove that there are infinitely many primes of the form $4k + 1$. Suppose that $p_1, p_2, \ldots, p_m$ are a finite set of such primes and consider $(2p_1p_2 \cdots p_m)^2 + 1$. Suppose that p divides this integer. -1 will then be a quadratic residue mod p and thus p will be of the form $4k + 1$. p is not among the p_i since $(2p_1p_2 \cdots p_m)^2 + 1$ leaves a remainder of 1 when divided by p_i. We have shown that every finite set of primes of the form $4k + 1$ excludes some primes of that form. Thus the set of such primes is infinite.

To return to the theory of quadratic residues, we are now going to introduce another characterization of the symbol (a/p) due to Gauss.

Consider $S = \{-(p-1)/2, -(p-3)/2, \ldots, -1, 1, 2, \ldots, (p-1)/2\}$. This is called the set of least residues mod p. If $p \nmid a$, let μ be the number of negative least residues of the integers $a, 2a, 3a, \ldots, ((p-1)/2)a$. For example, let $p = 7$ and $a = 4$. Then $(p-1)/2 = 3$, and $1 \cdot 4$, $2 \cdot 4$, and $3 \cdot 4$ are congruent to -3, 1, and -2, respectively. Thus in this case $\mu = 2$.

Lemma (Gauss's Lemma)

$(a/p) = (-1)^{\mu}$.

PROOF

Let $\pm m_l$ be the least residue of la, where m_l is positive. As l ranges between 1 and $(p-1)/2$, μ is clearly the number of minus signs that occur in this way. We claim that $m_l \neq m_k$ if $l \neq k$ and $1 \leq l, k \leq (p-1)/2$. For, if $m_l = m_k$, then $la \equiv \pm ka\ (p)$, and since $p \nmid a$ this implies that $l \pm k \equiv 0\ (p)$. The latter congruence is impossible since $l \neq k$ and $|l \pm k| \leq |l| + |k| \leq p - 1$. It follows that the sets $\{1, 2, \ldots, (p-1)/2\}$ and $\{m_1, m_2, \ldots, m_l\}$ coincide. Multiply the congruences $1 \cdot a \equiv \pm m_1\ (p)$, $2 \cdot a \equiv \pm m_2\ (p), \ldots, ((p-1)/2)a \equiv \pm m_{(p-1)/2}\ (p)$. We obtain

$$\left(\frac{p-1}{2}\right)! a^{(p-1)/2} \equiv (-1)^{\mu}\left(\frac{p-1}{2}\right)!\ (p)$$

This yields $a^{(p-1)/2} \equiv (-1)^{\mu}\ (p)$. By Proposition 5.1.2, $a^{(p-1)/2} \equiv (a/p)\ (p)$. The result follows.

Gauss's lemma is an extremely powerful tool. We shall base our first proof of the quadratic reciprocity law on it. Before getting to that, however, we can use it immediately to get a characterization of those primes for which 2 is a quadratic residue.

Proposition 5.1.3

2 is a quadratic residue of primes of the form $8k + 1$ *and* $8k + 7$. *2 is a quadratic nonresidue of primes of the form* $8k + 3$ *and* $8k + 5$. *This information is summarized in the formula*

$$(2/p) = (-1)^{(p^2-1)/8}$$

PROOF

We leave to the reader the task of showing that the formula is equivalent to the first two assertions.

Let p be an odd prime (as usual) and notice that the number μ is equal to the number of elements of the set $2 \cdot 1, 2 \cdot 2, \ldots, 2 \cdot (p-1)/2$ that exceed $(p-1)/2$. Let m be determined by the two conditions $2m \leq (p-1)/2$ and $2(m+1) > (p-1)/2$. Then $\mu = ((p-1)/2) - m$.

If $p = 8k + 1$, then $(p-1)/2 = 4k$ and $m = 2k$. Thus $\mu = 4k - 2k = 2k$ is even and $(2/p) = 1$.

If $p = 8k + 7$, then $(p-1)/2 = 4k + 3$, $m = 2k + 1$, and $\mu = 4k + 3 - (2k + 1) = 2k + 2$ is even. Thus $(2/p) = 1$ in this case as well.

If $p = 8k + 3$, then $(p-1)/2 = 4k + 1$, $m = 2k$, and $\mu = 4k + 1 - 2k = 2k + 1$ is odd. Thus $(2/p) = -1$.

Finally, if $p = 8k + 5$, then $(p-1)/2 = 4k + 2$, $m = 2k + 1$, and $\mu = 4k + 2 - (2k + 1) = 2k + 1$ is odd. Thus $(2/p) = -1$ and we are done.

As an example, consider $p = 7$ and $p = 17$. These primes are congruent to 7 and 1, respectively, mod 8, and indeed $3^2 \equiv 2$ (7) and $6^2 \equiv 2$ (17). On the other hand, $p = 19$ and $p = 5$ are congruent to 3 and 5, respectively, and it is easily checked numerically that 2 is a quadratic nonresidue for both primes.

One can use Proposition 5.1.3 to prove that there are infinitely many primes of the form $8k + 7$. Let $p_1, \ldots, p_m$ be a finite collection of such primes, and consider $(4p_1p_2 \cdots p_m)^2 - 2$. The odd prime divisors of this number have the form $8k + 1$ or $8k + 7$, since for such prime divisors 2 is a quadratic residue. Not all the odd prime divisors can have the form $8k + 1$ (prove it). Let p be a prime divisor of the form $8k + 7$. Then p is not in the set $\{p_1, p_2, \ldots, p_n\}$ and we are done.

2 LAW OF QUADRATIC RECIPROCITY

***Theorem 1** (**Law of Quadratic Reciprocity**)*

Let p and q be odd primes. Then

(a) $(-1/p) = (-1)^{(p-1)/2}$.

(b) $(2/p) = (-1)^{(p^2-1)/8}$.

(c) $(p/q)(q/p) = (-1)^{((p-1)/2)((q-1)/2)}$.

We are going to postpone the proof until Section 3. In Chapter 6 we shall prove the theorem once again from a different standpoint, and also indicate something of its history. It is among the deepest and most beautiful results of elementary number theory and the beginning of a line of reciprocity theorems that culminate in the very general Artin reciprocity law, perhaps the most impressive theorem in all number theory. It would take us far outside the compass of this book to even state the Artin reciprocity law, but in Chapter 9 we shall state and prove the law of cubic reciprocity.

Parts (a) and (b) of Theorem 1 have already been proved and some of their consequences discussed. Let us turn our attention to part (c).

If either p or q are of the form $4k + 1$, then $((p - 1)/2)((q - 1)/2) = 1$. If both p and q are of the form $4k + 3$, then $((p - 1)/2)((q - 1)/2) = -1$. This permits us to restate part (c) as follows:

1. If either p or q are of the form $4k + 1$, then p is a quadratic residue mod q iff q is a quadratic residue mod p.

2. If both p and q are of the form $4k + 3$, then p is a quadratic residue mod q iff q is a quadratic nonresidue mod p.

As a first application of quadratic reciprocity we show how, in conjunction with Proposition 5.1.2, it can be used in numerical computations of the Legendre symbol. A single example should suffice to illustrate the method.

We propose to calculate $(79/101)$. Since $101 \equiv 1\ (4)$ we have $(79/101) = (101/79) = (22/79)$. The last step follows from $101 \equiv 22\,(79)$. Further, $(22/79) = (2/79)(11/79)$. Now $79 \equiv 7\ (8)$. Thus $(2/79) = 1$. Since both 11 and 79 are congruent to 3 mod 4 we have $(11/79) = -(79/11) = -(2/11)$. Finally $11 \equiv 3\ (8)$ implies that $(2/11) = -1$. Therefore $(79/101) = 1$; i.e., 79 is a quadratic residue mod 101. Indeed, $33^2 \equiv 79\,(101)$.

The next application is perhaps more significant. We noticed earlier that -1 is a quadratic residue of primes of the form $4k + 1$ and that 2 is a quadratic residue of primes that are either of the form $8k + 1$ or $8k + 7$. If a is an arbitrary integer, for what primes p is a a quadratic residue mod p? We are now in a position to give an answer. To begin with, we consider the case where $a = q$, an odd prime.

Theorem 2

Let q be an odd prime.

(a) *If $q \equiv 1\ (4)$, then q is a quadratic residue* mod *p iff $p \equiv r\ (q)$, where r is a quadratic residue* mod *q.*

(b) *If $q \equiv 3\ (4)$, then q is a quadratic residue* mod *p iff $p \equiv \pm b^2\ (4q)$, where b is an odd integer prime to q.*

PROOF

If $q \equiv 1\ (4)$, then by Theorem 1 we have $(q/p) = (p/q)$. Part (a) is thus clear.

If $q \equiv 3\ (4)$, Theorem 1 yields $(q/p) = (-1)^{(p-1)/2}(p/q)$. Assume first that $p \equiv \pm b^2\ (4q)$, where b is odd. If we take the plus sign, we get $p \equiv b^2 \equiv 1\ (4)$ and $p \equiv b^2\ (q)$. Thus $(-1)^{(p-1)/2} = 1$ and $(p/q) = 1$, giving $(q/p) = 1$. If we take the minus sign, then $p \equiv -b^2 \equiv -1 \equiv 3\ (4)$ and $p \equiv -b^2\ (q)$. The first congruence shows that $(-1)^{(p-1)/2} = -1$. The second shows that $(p/q) = (-b^2/q) = (-1/q)(b/q)^2 = (-1/q) = -1$ since $q \equiv 3\ (4)$. Once again we have $(q/p) = 1$.

To go the other way, assume that $(q/p) = 1$. We have two cases to deal with:

1. $(-1)^{(p-1)/2} = -1$ and $(p/q) = -1$.
2. $(-1)^{(p-1)/2} = 1$ and $(p/q) = 1$.

In case 2 we have $p \equiv b^2(q)$ and $p \equiv 1\ (4)$. b can be assumed to be odd since if it is even we can use $b' = b + q$ instead. If b is odd, then $b^2 \equiv 1\ (4)$ and $p \equiv b^2\ (4)$ and thus $p \equiv b^2\ (4q)$, as required.

In case 1 we have $p \equiv 3\ (4)$ and $p \equiv -b^2\ (q)$. The last congruence follows since $q \equiv 3\ (4)$ implies that every nonresidue is the negative of a residue (prove it). Again, we may assume that b is odd. In that case

$-b^2 \equiv 3\ (4)$ so $p \equiv -b^2\ (4)$ and $p \equiv -b^2\ (4q)$. This concludes the proof.

Take $q = 3$ as a first illustration. By part (b) of Theorem 2 we must find the residues mod 12 of the squares of odd integers prime to 3. $1^2, 5^2, 7^2$, and 11^2 are all congruent to 1. Thus 3 is a quadratic residue of primes p congruent to $\pm 1\ (12)$ and a quadratic nonresidue of primes congruent to $\pm 5\ (12)$.

Next consider $q = 5$. Since $5 \equiv 1\ (4)$ we are in the simpler part (a) of Theorem 2. 1 and 4 are the residues mod 5, and 2 and 3 the non-residues. Thus 5 is a residue of primes congruent to 1 or 4 mod 5 and a nonresidue of primes congruent to 2 or 3 mod 5.

"Numbers congruent to b mod m" and "numbers of the form $mk + b$" are short-hand expressions describing the set $\{b, b + m, b + 2m, \ldots\}$. This set is an arithmetic progression with initial term b and difference m. In our investigations so far we have seen that the answer to the question for which primes p is a a quadratic residue has been for those primes p that occur in a certain fixed, finite number of arithmetic progressions. This situation is entirely general. Instead of stating this result as a theorem (the statement would be very complicated) we shall work out a few numerical examples.

For $a = -3$, $(-3/p) = (-1/p)(3/p)$. Thus -3 is a quadratic residue mod p if either $(-1/p) = 1$ and $(3/p) = 1$ or $(-1/p) = -1$ and $(3/p) = -1$.

By our previous results the first case obtains when $p \equiv 1\ (4)$ and $p \equiv \pm 1\ (12)$. If $p \equiv -1\ (12)$, then $p \equiv -1\ (4)$. The only primes that satisfy both congruences are $\equiv 1\ (12)$.

In the second case $p \equiv 3\ (4)$ and $p \equiv \pm 5\ (12)$. If $p \equiv 5\ (12)$, then $p \equiv 1\ (4)$. Thus the only primes that satisfy both these congruences are $\equiv -5\ (12)$.

Summarizing, -3 is a quadratic residue mod p iff p is congruent to 1 or -5 mod 12.

Now consider $a = 6$. Since $(6/p) = (2/p)(3/p)$ we again have two cases: $(2/p) = 1$ and $(3/p) = 1$ or $(2/p) = -1$ and $(3/p) = -1$. The first case holds if $p \equiv 1, 7\ (8)$ and $p \equiv 1, 11\ (12)$. The only two pairs of congruences that are compatible are $p \equiv 1\ (8)$ and $p \equiv 1\ (12)$, and $p \equiv 7\ (8)$ and $p \equiv 11\ (12)$. By standard techniques (see Chapter 3) the primes satisfying these congruences are congruent to 1 or 23 mod 24.

In the second case we have to consider $p \equiv 3, 5\ (8)$ and $p \equiv 5, 7\ (12)$. Separating these into four pairs of congruences we see that the only solutions are congruent to 5 and 19 mod 24.

Summarizing, 6 is a quadratic residue mod p iff $p \equiv 1, 5, 19, 23\ (24)$.

As a numerical check we see for the primes 73, 5, 19, and 23 that $15^2 \equiv 6\ (73)$, $1^2 \equiv 6\ (5)$, $5^2 \equiv 6\ (19)$, and $11^2 \equiv 6\ (23)$.

As a final application of the quadratic reciprocity law we investigate the question. If a is a quadratic residue mod all primes p not dividing a, what can be said about a? If a is a square, it is a residue for all primes not dividing a. It turns out that the converse of this statement is true as well. In fact, we shall soon prove an even stronger result. First, however, it is necessary to define and investigate briefly a new symbol.

Definition

Let b be an odd, positive integer and a any integer. Let $b = p_1 p_2 \cdots p_m$, where the p_i are (not necessarily distinct) primes. The symbol (a/b) defined by

$$(a/b) = (a/p_1)(a/p_2)\cdots(a/p_m)$$

is called the *Jacobi symbol*.

The Jacobi symbol has properties that are remarkably similar to the Legendre symbol, which it generalizes. A word of caution is useful. (a/b) may equal 1 without a being a quadratic residue mod b. For example, $(2/15) = (2/3)(2/5) = (-1)(-1) = 1$, but 2 is not a residue mod 15. It is true, however, that if $(a/b) = -1$, then a is a nonresidue mod b.

Proposition 5.2.1

(a) $(a_1/b) = (a_2/b)$ if $a_1 \equiv a_2$ (b).
(b) $(a_1 a_2/b) = (a_1/b)(a_2/b)$.
(c) $(a/b_1 b_2) = (a/b_1)(a/b_2)$.

PROOF

Parts (a) and (b) are immediate from the corresponding properties of the Legendre symbol. Part (c) is obvious from the definition.

Lemma

Let r and s be odd integers. Then

(a) $(rs - 1)/2 \equiv ((r - 1)/2) + ((s - 1)/2)$ (2).
(b) $(r^2 s^2 - 1)/8 \equiv ((r^2 - 1)/8) + ((s^2 - 1)/8)$ (2).

PROOF

Since $(r - 1)(s - 1) \equiv 0$ (4) we have $rs - 1 \equiv (r - 1) + (s - 1)$ (4). Part (a) follows by dividing by 2.

$r^2 - 1$ and $s^2 - 1$ are both divisible by 4. Thus $(r^2 - 1)(s^2 - 1) \equiv 0$ (16) and $r^2 s^2 - 1 \equiv (r^2 - 1) + (s^2 - 1)$ (16). Part (b) follows upon dividing by 8.

Corollary

Let $r_1, r_2, \ldots, r_m$ be odd integers. Then

(a) $\sum_{i=1}^{m} (r_i - 1)/2 \equiv (r_1 r_2 \cdots r_m - 1)/2$ (2).

(b) $\sum_{i=1}^{m} (r_i^2 - 1)/8 \equiv (r_1^2 r_2^2 \cdots r_m^2 - 1)/8$ (2).

PROOF

The proof is a simple induction on m, using the lemma.

Proposition 5.2.2

(a) $(-1/b) = (-1)^{(b-1)/2}$.

(b) $(2/b) = (-1)^{(b^2-1)/8}$.

(c) *If a is odd and positive as well as b, then*

$$(a/b)(b/a) = (-1)^{((a-1)/2)((b-1)/2)}$$

PROOF

$$(-1/b) = (-1/p_1)(-1/p_2)\cdots(-1/p_m) = (-1)^{(p_1-1)/2} \cdots (-1)^{(p_m-1)/2} = (-1)^{\Sigma(p_i-1)/2}.$$

By the lemma $\sum (p_i - 1)/2 \equiv (p_1 p_2 \cdots p_m - 1)/2 \equiv (b - 1)/2$ (2). This proves part (a).

Part (b) is proved in exactly the same way.

Now if $a = q_1 q_2 \cdots q_l$, then

$$(a/b)(b/a) = \prod_i \prod_j (q_i/p_j)(p_j/q_i) = (-1)^{\Sigma_i \Sigma_j ((q_i-1)/2)((p_j-1)/2)}$$

The product and sum range over $1 \le i \le l$ and $1 \le j \le m$. Again using the lemma we have

$$\sum_i \sum_j ((p_j - 1)/2)((q_i - 1)/2) \equiv (a - 1)/2 \sum_i (p_i - 1)/2$$
$$\equiv ((a - 1)/2)((b - 1)/2) \ (2)$$

This proves part (c).

The Jacobi symbol has many uses. For one thing, it is a convenient aid for computing the Legendre symbol. We now use it to prove the following theorem.

Theorem 3

Let a be a nonsquare integer. Then there are infinitely many primes p for which a is a quadratic nonresidue.

PROOF

It is easily seen that we may assume that a is square-free. Let $a = 2^e q_1 q_2 \cdots q_n$, where the q_i are distinct odd primes and $e = 0$ or 1. The case $a = 2$ has to be dealt with separately. We shall assume to begin with that $n \ge 1$, i.e., that a is divisible by an odd prime.

Let $l_1, l_2, \ldots, l_k$ be a finite set of odd primes not including any q_i. Let s be any nonresidue mod q_n, and find a simultaneous solution to the congruences

$$x \equiv 1\ (l_i) \qquad i = 1, \ldots, k$$

$$x \equiv 1\ (8)$$

$$x \equiv 1\ (q_i) \qquad i = 1, 2, \ldots, n-1$$

$$x \equiv s\ (q_n)$$

Call the solution b. b is odd. Suppose that $b = p_1 p_2 \cdots p_m$ is its prime decomposition. Since $b \equiv 1$ (8) we have $(2/b) = 1$ and $(q_i/b) = (b/q_i)$ by Proposition 5.2.2. Thus $(a/b) = (2/b)^e(q_1/b) \cdots (q_{n-1}/b)(q_n/b) = (b/q_1) \cdots (b/q_{n-1})(b/q_n) = (1/q_1) \cdots (1/q_{n-1})(s/q_n) = -1$.

On the other hand, by the definition of (a/b), we have $(a/b) = (a/p_1)(a/p_2) \cdots (a/p_m)$. It follows that $(a/p_i) = -1$ for some i.

Notice that l_j does not divide b. Thus $p_i \notin \{l_1, l_2, \ldots, l_k\}$.

To summarize, if a is a nonsquare, divisible by an odd prime, we have found a prime p, outside a given finite set of primes $\{2, l_1, l_2, \ldots, l_k\}$, such that $(a/p) = -1$. This proves Theorem 3 in this case.

It remains to consider the case $a = 2$. Let $l_1, \ldots, l_k$ be a finite set of primes, excluding 3, for which $(2/l_i) = -1$. Let $b = 8l_1 l_2 \cdots l_k + 3$. b is not divisible by 3 or any l_i. Since $b \equiv 3$ (8) we have $(2/b) = (-1)^{(b^2-1)/8} = -1$. Suppose that $b = p_1 p_2 \cdots p_m$ is the prime decomposition of b. Then, as before, we see that $(2/p_i) = -1$ for some i. $p_i \notin \{3, l_1, l_2, \ldots, l_k\}$. This proves Theorem 3 for $a = 2$.

3 A PROOF OF THE LAW OF QUADRATIC RECIPROCITY

Gauss found eight separate proofs for the law of quadratic reciprocity. There are over a hundred now in existence. Of course, they are not all essentially different. Many just differ in small details from others. We shall present an ingenious proof due to Eisenstein. For a somewhat more elementary and standard proof, see [61].

A complex number ζ is called an nth root of unity if $\zeta^n = 1$ for some integer $n > 0$. If n is the least integer with this property, then ζ is called a primitive nth root of unity.

The nth roots of unity are $1, e^{2\pi i/n}, e^{(2\pi i/n)2}, \ldots, e^{(2\pi i/n)(n-1)}$. Among these the primitive nth roots of unity are $e^{(2\pi i/n)k}$, where $(k, n) = 1$.

If ζ is an nth root of unity and $m \equiv l\ (n)$, then $\zeta^m = \zeta^l$. If ζ is a primitive nth root of unity and $\zeta^m = \zeta^l$, then $m \equiv l\ (n)$.

These elementary properties are easy to prove.

Consider the function $f(z) = e^{2\pi iz} - e^{-2\pi iz} = 2i \sin 2\pi z$. This function satisfies $f(z+1) = f(z)$ and $f(-z) = -f(z)$. Also, its only real zeros are the integers. In other words, if r is a real number and $r \notin Z$, then $f(r) \neq 0$.

We wish to prove an important identity involving $f(z)$, but first we need an algebraic lemma.

Lemma

If $n > 0$ is odd, we have

$$x^n - y^n = \prod_{k=0}^{n-1} (\zeta^k x - \zeta^{-k} y), \qquad \textit{where } \zeta = e^{2\pi i/n}$$

PROOF

$1, \zeta, \zeta^2, \ldots, \zeta^{n-1}$ are all roots of the polynomial $z^n - 1$. Since there are n of them and they are all distinct we have $z^n - 1 = \prod_{k=0}^{n-1} (z - \zeta^k)$. Let $z = x/y$ and multiply both sides by y^n. We get $x^n - y^n = \prod_{k=0}^{n-1} (x^n - \zeta^k y^n)$.

Since n is odd as k runs over a complete system of residues mod n, so does $-2k$. Thus

$$\begin{aligned} x^n - y^n &= \prod_{k=0}^{n-1} (x - \zeta^{-2k} y) \\ &= \zeta^{-(1+2+\cdots+n-1)} \prod_{k=0}^{n-1} (\zeta^k x - \zeta^{-k} y) \\ &= \prod_{k=0}^{n-1} (\zeta^k x - \zeta^{-k} y) \end{aligned}$$

In the last step we have used the fact that $1 + 2 + 3 + \cdots + (n-1) = n((n-1)/2)$ is divisible by n.

Proposition 5.3.1

If n is a positive odd integer and $f(z) = e^{2\pi iz} - e^{-2\pi iz}$, then

$$\frac{f(nz)}{f(z)} = \prod_{k=1}^{(n-1)/2} f\left(z + \frac{k}{n}\right) f\left(z - \frac{k}{n}\right)$$

PROOF

In the lemma, substitute $x = e^{2\pi iz}$ and $y = e^{-2\pi iz}$. We see that

$$f(nz) = \prod_{k=0}^{n-1} f\left(z + \frac{k}{n}\right)$$

Notice that $f(z + k/n) = f(z + k/n - 1) = f(z - (n-k)/n)$. As k

goes from $(n+1)/2$ to $n-1$, $n-k$ goes from $(n-1)/2$ to 1. Thus

$$\begin{aligned}\frac{f(nz)}{f(z)} &= \prod_{k=1}^{(n-1)/2} f\left(z+\frac{k}{n}\right) \prod_{k=(n+1)/2}^{n-1} f\left(z+\frac{k}{n}\right)\\ &= \prod_{k=1}^{(n-1)/2} f\left(z+\frac{k}{n}\right) \prod_{k=(n+1)/2}^{n-1} f\left(z-\frac{n-k}{n}\right)\\ &= \prod_{k=1}^{(n-1)/2} f\left(z+\frac{k}{n}\right) f\left(z-\frac{k}{n}\right)\end{aligned}$$

Proposition 5.3.2

If p is an odd prime, $a \in \mathbb{Z}$, and $p \nmid a$, then

$$\prod_{l=1}^{(p-1)/2} f\left(\frac{la}{p}\right) = (a/p) \prod_{l=1}^{(p-1)/2} f\left(\frac{l}{p}\right)$$

PROOF

As in the lemma of Section 1, $la \equiv \pm m_l\ (p)$, where $1 \le m_l \le (p-1)/2$. Thus la/p and $\pm m_l/p$ differ by an integer. This implies that $f(la/p) = f(\pm m_l/p) = \pm f(m_l/p)$.

The result now follows by taking the product of both sides as l goes from 1 to $(p-1)/2$ and applying Gauss's lemma.

We are now in a position to prove the law of quadratic reciprocity. Let p and q be odd primes. Then by Proposition 5.3.2

$$\prod_{l=1}^{(p-1)/2} f\left(\frac{lq}{p}\right) = (q/p) \prod_{l=1}^{(p-1)/2} f\left(\frac{l}{p}\right)$$

By Proposition 5.3.1

$$\frac{f(ql/p)}{f(l/p)} = \prod_{m=1}^{(q-1)/2} f\left(\frac{l}{p}+\frac{m}{q}\right) f\left(\frac{l}{p}-\frac{m}{q}\right)$$

Putting these two equations together we have

$$(q/p) = \prod_{m=1}^{(q-1)/2} \prod_{l=1}^{(p-1)/2} f\left(\frac{l}{p}+\frac{m}{q}\right) f\left(\frac{l}{p}-\frac{m}{q}\right)$$

In the same way we find

$$(p/q) = \prod_{m=1}^{(q-1)/2} \prod_{l=1}^{(p-1)/2} f\left(\frac{m}{q}+\frac{l}{p}\right) f\left(\frac{m}{q}-\frac{l}{p}\right)$$

Since $f(m/q - l/p) = -f(l/p - m/q)$ we see that

$$(-1)^{((p-1)/2)((q-1)/2)}(q/p) = (q/p)$$

and therefore that

$$(p/q)(q/p) = (-1)^{((p-1)/2)((q-1)/2)}$$

The proof is complete.

Notes

It is generally believed that Legendre was the first to state the law of quadratic reciprocity in 1785. However, equivalent formulations are found in the works of Euler and, in fact, special cases of reciprocity are easily deduced from the letters of Fermat (for example, that 5 is a quadratic residue modulo all primes of the form $5t + 1$). Gauss is indisputably the first to offer a complete proof. He found eight proofs in all. Other proofs have followed throughout the nineteenth and twentieth centuries. Among other prominent names, G. Eisenstein, A. Cauchy, R. Dedekind, Dirichlet, and Kronecker have made contributions. In 1921 there were 56 known proofs. M. Gerstenhaber published a paper in 1963 that he claimed contained the one hundred fifty-second proof of quadratic reciprocity. One deduces that at least 96 additional proofs have come out between 1921 and 1963. It is a question of some psychological interest why mathematicians are so obsessed with proving this theorem. For a discussion of Gauss's various proofs, see Rieger [66], as well as Bachman's erudite essay [6]. See also the interesting proof of R. Swan [75].

The Jacobi symbol is one generalization of the Legendre symbol. For an interesting generalization in another direction, see the paper of P. Cartier [14].

Quadratic reciprocity can be formulated in rings other than Z. Dirichlet proved such a theorem for the ring of Gaussian integers $Z[i]$. D. Hilbert was able to prove that quadratic reciprocity held for any algebraic number field, a result that was an important stepping stone to class field theory. In another direction it can be shown that reciprocity holds for the ring $k[x]$, where k is a finite field. See Artin [2] and Carlitz [10]. This result had already been stated (though not proved) by Dedekind in 1857.

Theorem 3 has been completed by N. C. Ankeny and C. A. Rogers (*Annals of Math.*, **53**, 1951, 541–550). They prove that if the congruence $x^n \equiv a\ (p)$ has a solution for all but finitely many primes p, then either $a = b^n$ or $8 \mid n$ and $a = 2^{n/2}b^n$.

Exercises

1 Use Gauss's lemma to determine (5/7), (3/11), (6/13), and $(-1/p)$.

2 Show that the number of solutions to $x^2 \equiv a\ (p)$ is given by $1 + (a/p)$.

3 Suppose that $p \nmid a$. Show that the number of solutions to $ax^2 + bx + c \equiv 0\ (p)$

is given by $1 + ((b^2 - 4ac)/p)$.

4 Prove that $\sum_{a=1}^{p-1} (a/p) = 0$.

5 Prove that $\sum_{x=1}^{p-1} ((ax + b)/p) = 0$ provided that $p \not| a$.

6 Show that the number of solutions to $x^2 - y^2 \equiv a\,(p)$ is given by $\sum_{y=0}^{p-1} (1 + ((y^2 + a)/p))$.

7 By calculating directly show that the number of solutions to $x^2 - y^2 \equiv a\,(p)$ is $p - 1$ if $p \not| a$ and $2p - 1$ if $p \mid a$. (*Hint:* Use the change of variables $u = x + y$, $v = x - y$.)

8 Combining the results of Exercises 6 and 7 show that

$$\sum_{y=0}^{p-1} ((y^2 + a)/p) = \begin{cases} -1 & \text{if } p \not| a \\ p - 1 & \text{if } p \mid a \end{cases}$$

9 Prove that $1^2 3^2 5^2 \cdots (p - 2)^2 \equiv (-1)^{(p+1)/2}\,(p)$ by using Wilson's theorem.

10 Let $r_1, r_2, \ldots, r_{(p-1)/2}$ be the quadratic residues between 1 and p. Show that their product is congruent to $1\,(p)$ if $p \equiv 3\,(4)$ and congruent to $-1\,(p)$ if $p \equiv 1\,(4)$.

11 Suppose that $p \equiv 3\,(4)$ and that $q \equiv 2p + 1$ is also prime. Prove that $2^p - 1$ is not prime. (*Hint:* Use the quadratic character of 2 to show that $q \mid 2^p - 1$.) One must assume that $p > 3$.

12 Let $f(x) \in \mathbb{Z}[x]$. We say that a prime p divides $f(x)$ if there is an integer n such that $p \mid f(n)$. Describe the prime divisors of $x^2 + 1$ and $x^2 - 2$.

13 Show that any prime divisor of $x^4 - x^2 + 1$ is congruent to 1 modulo 12.

14 Use the fact that $U(\mathbb{Z}/p\mathbb{Z})$ is cyclic to give a direct proof that $(-3/p) = 1$ when $p \equiv 1\,(3)$. [*Hint:* There is a ρ in $U(\mathbb{Z}/p\mathbb{Z})$ of order 3. Show that $(2\rho + 1)^2 = -\bar{3}$.]

15 If $p \equiv 1\,(5)$, show directly that $(5/p) = 1$ by the method of Exercise 14. [*Hint:* Let ρ be an element of $U(\mathbb{Z}/p\mathbb{Z})$ of order 5. Show that $(\rho + \rho^4)^2 + (\rho + \rho^4) - \bar{1} = \bar{0}$, etc.]

16 Using quadratic reciprocity find the primes for which 7 is a quadratic residue. Do the same for 15.

17 Supply the details to the proof of Proposition 5.2.1 and to the corollary to the lemma following it.

18 Let D be a square-free integer that is also odd and positive. Show that there is an integer b prime to D such that $(b/D) = -1$.

19 Let D be as in Exercise 18. Show that $\sum (a/D) = 0$, where the sum is over a reduced residue system modulo D (see Exercise 6 of Chapter 3). Conclude that exactly one half of the elements in $U(\mathbb{Z}/D\mathbb{Z})$ satisfy $(a/D) = 1$.

20 (continuation) Let $a_1, a_2, \ldots, a_{\phi(D)/2}$ be integers between 1 and D such that $(a_i, D) = 1$ and $(a_i/D) = 1$. Prove that D is a quadratic residue modulo a prime p iff $p \equiv a_i\,(D)$ for some i.

21 Apply the method of Exercises 19 and 20 to find those primes for which 21 is a quadratic residue. [*Answer:* Those $p \equiv 1, 4, 5, 16, 17$, and $20\,(21)$.]

22 Use the Jacobi symbol to determine (113/997), (215/761), (514/1093), and (401/757).

23 Suppose that $p \equiv 1\,(4)$. Show that there exist integers s and t such that $pt = 1 + s^2$. Conclude that p is not a prime in $\mathbb{Z}[i]$. Remember that $\mathbb{Z}[i]$ has unique factorization.

24 If $p \equiv 1\ (4)$, show that p is the sum of two squares; i.e., $p = a^2 + b^2$ with $a, b \in \mathbb{Z}$. (*Hint:* $p = \alpha\beta$ with α and β being nonunits in $\mathbb{Z}[i]$. Take the absolute value of both sides and square the result.) This important result was discovered by Fermat.

25 An integer is called a biquadratic residue modulo p if it is congruent to a fourth power. Using the identity $x^4 + 4 = ((x + 1)^2 + 1)((x - 1)^2 + 1)$ show that -4 is a biquadratic residue modulo p iff $p \equiv 1\ (4)$.

26 This exercise and Exercises 27 and 28 give Dirichlet's beautiful proof that 2 is a biquadratic residue modulo p iff p can be written in the form $A^2 + 64B^2$, where $A, B \in \mathbb{Z}$. Suppose that $p \equiv 1\ (4)$. Then $p = a^2 + b^2$ by Exercise 24. Take a to be odd. Prove the following statements:

(a) $(a/p) = 1$.
(b) $((a + b)/p) = (-1)^{((a+b)^2-1)/8}$.
(c) $(a + b)^2 \equiv 2ab\ (p)$.
(d) $(a + b)^{(p-1)/2} \equiv (2ab)^{(p-1)/4}\ (p)$.

[*Hint:* $2p = (a + b)^2 + (a - b)^2$.]

27 Suppose that f is such that $b \equiv af\ (p)$. Show that $f^2 \equiv -1\ (p)$ and that $2^{(p-1)/4} \equiv f^{ab/2}\ (p)$.

28 Show that $x^4 \equiv 2\ (p)$ has a solution iff p is of the form $A^2 + 64B^2$.

29 Let (RR) be the number of pairs $(n, n + 1)$ in the set $1, 2, 3, \ldots, p - 1$ such that n and $n + 1$ are both quadratic residues modulo p. Let (NR) be the number of pairs $(n, n + 1)$ in the set $1, 2, 3, \ldots, p - 1$ such that n is a quadratic nonresidue and $n + 1$ is a quadratic residue. Similarly, define (RN) and (NN). Determine the sums $(RR) + (RN)$, $(NR) + (NN)$, $(RR) + (NR)$, and $(RN) + (NN)$.

30 Show that $(RR) + (NN) - (RN) - (NR) = \sum_{n=1}^{p-1} (n(n + 1))/p$. Evaluate this sum and show that it is equal to -1. (*Hint:* The result of Exercise 8 is useful.)

31 Use the results of Exercises 29 and 30 to show that $(RR) = \frac{1}{4}(p - 4 - \varepsilon)$, where $\varepsilon = (-1)^{(p-1)/2}$.

chapter six/QUADRATIC GAUSS SUMS

The method by which we proved the quadratic reciprocity in Chapter 5 is ingenious but is not easy to use in more general situations. We shall give a new proof in this chapter that is based on methods that can be used to prove higher reciprocity laws. In particular, we shall introduce the notion of a Gauss sum, which will play an important role in the latter part of this book.

Section 1 introduces algebraic numbers and algebraic integers. The proofs are somewhat technical. The reader may wish to simply skim this section on a first reading.

I ALGEBRAIC NUMBERS AND ALGEBRAIC INTEGERS

Definition

An *algebraic number* is a complex number α that is a root of a polynomial $a_0x^n + a_1x^{n-1} + a_2x^{n-2} + \cdots + a_n = 0$, where $a_0, a_1, a_2, \ldots, a_n \in \mathbb{Q}$. and $a_0 \neq 0$.

An *algebraic integer* ω is a complex number that is a root of a polynomial $x^n + b_1x^{n-1} + \cdots + b_n = 0$, where $b_1, b_2, \ldots, b_n \in \mathbb{Z}$.

Clearly every algebraic integer is an algebraic number. The converse is false, as we shall see.

Proposition 6.1.1

A rational number $r \in \mathbb{Q}$ is an algebraic integer iff $r \in \mathbb{Z}$.

PROOF

If $r \in \mathbb{Z}$, then r is a root of $x - r = 0$. Thus r is an algebraic integer.

Suppose that $r \in \mathbb{Q}$ and that r is an algebraic integer; i.e., r satisfies an equation $x^n + b_1x^{n-1} + \cdots + b_n = 0$ with $b_1, \ldots, b_n \in \mathbb{Z}$. $r = c/d$, where $c, d \in \mathbb{Z}$ and we may assume that c and d are relatively prime. Substituting c/d into the equation and multiplying both sides by d^n yields

$$c^n + b_1c^{n-1}d + \cdots + b_nd^n = 0$$

It follows that d divides c^n and, since $(d, c) = 1$, that $d \mid c$. Again, since $(d, c) = 1$ it follows that $d = \pm 1$, and so $r = c/d$ is in $\mathbb{Z}$.

It follows, for example, that $\frac{2}{5}$ is not an algebraic integer.

The main results of this section are that the set of algebraic numbers form a field and that the set of algebraic integers form a ring. We need some preliminary work.

Definition

A *subset* $V \subset \mathbb{C}$ of the complex numbers is called a $\mathbb{Q}$ module if

(a) $\gamma_1, \gamma_2 \in V$ implies that $\gamma_1 \pm \gamma_2 \in V$.

(b) $\gamma \in V$ and $r \in \mathbb{Q}$ implies that $r\gamma \in V$.

(c) There exist elements $\gamma_1, \gamma_2, \ldots, \gamma_l \in V$ such that every $\gamma \in V$ has the form $\sum_{i=1}^{l} r_i\gamma_i$ with $r_i \in \mathbb{Q}$.

More briefly, $V \subset \mathbb{C}$ is a $\mathbb{Q}$ module if it is a finite dimensional vector space over $\mathbb{Q}$.

If $\gamma_1, \gamma_2, \ldots, \gamma_l \in \mathbb{C}$, the set of all expressions $\sum_{i=1}^{l} r_i\gamma_i, r_1, r_2, \ldots, r_l \in \mathbb{Q}$ is easily seen to be a $\mathbb{Q}$ module. We denote this $\mathbb{Q}$ module by $[\gamma_1, \gamma_2, \ldots, \gamma_l]$.

Proposition 6.1.2

Let $V = [\gamma_1, \gamma_2, \ldots, \gamma_l]$, and suppose that $\alpha \in \mathbb{C}$ has the property that $\alpha\gamma \in V$ for all $\gamma \in V$. Then α is an algebraic number.

PROOF

$\alpha\gamma_i \in V$ for $i = 1, 2, \ldots, l$. Thus $\alpha\gamma_i = \sum_{i=1}^{l} a_{ij}\gamma_j$, where $a_{ij} \in \mathbb{Q}$. It follows that $0 = \sum_{i=1}^{l} (a_{ij} - \delta_{ij}\alpha)\gamma_j$, where $\delta_{ij} = 0$ if $i \neq j$ and $\delta_{ij} = 1$ if $i = j$. By standard linear algebra we have that $\det(a_{ij} - \delta_{ij}\alpha) = 0$. Writing out the determinant we see that α satisfies a polynomial of degree l with rational coefficients. Thus α is an algebraic number.

Proposition 6.1.3

The set of algebraic numbers forms a field.

PROOF

Suppose that α_1 and α_2 are algebraic numbers. We shall show that $\alpha_1\alpha_2$ and $\alpha_1 + \alpha_2$ are algebraic numbers.

Suppose that $\alpha_1^n + r_1\alpha_1^{n-1} + r_2\alpha_1^{n-2} + \cdots + r_n = 0$ and that $\alpha_2^m + s_1\alpha_2^{m-1} + s_2\alpha_2^{m-2} + \cdots + s_m = 0$, where $r_i, s_j \in \mathbb{Q}$. Let V be the $\mathbb{Q}$ module obtained by forming all $\mathbb{Q}$ linear combinations of the elements $\alpha_1^i\alpha_2^j$, where $0 \leq i < n$ and $0 \leq j < m$. For $\gamma \in V$ we have $\alpha_1\gamma \in V$ and $\alpha_2\gamma \in V$ (prove it). Thus we also have $(\alpha_1 + \alpha_2)\gamma \in V$ and $(\alpha_1\alpha_2)\gamma \in V$. By Proposition 6.1.3 it follows that both $\alpha_1 + \alpha_2$ and $\alpha_1\alpha_2$ are algebraic numbers.

Finally, if α is an algebraic number, not zero, we must show that α^{-1} is an algebraic number. Suppose that $a_0\alpha^n + a_1\alpha^{n-1} + \cdots + a_n = 0$, where the $a_i \in \mathbb{Q}$. Then $a_n\alpha^{-n} + a_{n-1}\alpha^{-(n-1)} + \cdots + a_0 = 0$. The result follows.

To prove that the set of algebraic integers form a ring it is necessary only to alter the above proofs slightly.

Definition

A subset $W \subset \mathbb{C}$ is called a $\mathbb{Z}$ *module* if

(a) $\gamma_1, \gamma_2 \in W$ implies that $\gamma_1 \pm \gamma_2 \in W$.

(b) There exist elements $\gamma_1, \gamma_2, \ldots, \gamma_l \in W$ such that every $\gamma \in W$ is of the form $\sum_{i=1}^{l} b_i\gamma_i$ with $b_i \in \mathbb{Z}$.

Proposition 6.1.4

Let W be a $\mathbb{Z}$ module and suppose that $\omega \in \mathbb{C}$ is such that $\omega\gamma \in W$ for all $\gamma \in W$. Then ω is an algebraic integer.

PROOF

The proof proceeds exactly as in Proposition 6.1.2, except that now the $a_{ij} \in \mathbb{Z}$. The equation $\det(a_{ij} - \delta_{ij}\omega) = 0$ when written out shows that ω satisfies a monic equation of degree l with integer coefficients. Thus ω is an algebraic integer.

Proposition 6.1.5

The set of algebraic integers form a ring.

PROOF

The proof follows from Proposition 6.1.4 in exactly the same way in which Proposition 6.1.3 follows from Proposition 6.1.2. We leave the details to the reader.

Let Ω denote the ring of algebraic integers. If $\omega_1, \omega_2, \gamma \in \Omega$, we say that $\omega_1 \equiv \omega_2\,(\gamma)$ (ω_1 is congruent to ω_2 modulo γ) if $\omega_1 - \omega_2 = \gamma\alpha$ with $\alpha \in \Omega$. This notion of congruence satisfies all the formal properties of congruence in $\mathbb{Z}$.

If $a, b, c \in \mathbb{Z}$, $c \neq 0$, then $a \equiv b\,(c)$ is ambiguous since it denotes congruence in $\mathbb{Z}$ and in Ω. The ambiguity is only apparent, however. If $a - b = c\alpha$ with $\alpha \in \Omega$, then α is both a rational number and an algebraic integer. Thus α is an ordinary integer by Proposition 6.1.1.

The following proposition will be useful.

Proposition 6.1.6

If $\omega_1, \omega_2 \in \Omega$ and $p \in \mathbb{Z}$ is a prime, then

$$(\omega_1 + \omega_2)^p \equiv \omega_1^p + \omega_2^p\,(p)$$

PROOF

$(\omega_1 + \omega_2)^p = \sum_{k=0}^{p} \binom{p}{k} \omega_1^k \omega_2^{p-k}$. By Lemma 2 of Chapter 4 we have $p \mid \binom{p}{k}$ for $1 \leq k \leq p - 1$. The result follows from this and the fact that Ω is a ring.

One final remark. A root of unity is a solution to an equation of the form $x^n - 1 = 0$. Thus roots of unity are algebraic integers, and so are $\mathbb{Z}$ linear combinations of roots of unity.

2 *THE QUADRATIC CHARACTER OF 2*

Let $\zeta = e^{2\pi i/8}$. Then ζ is a primitive eighth root of unity. Thus $0 = \zeta^8 - 1 = (\zeta^4 - 1)(\zeta^4 + 1)$. Since $\zeta^4 \neq 1$ we have $\zeta^4 = -1$. Multiplying by ζ^{-2} and then adding ζ^{-2} to both sides yields $\zeta^2 + \zeta^{-1} = 0$. This equation is also easily derived from the observation that $\zeta^2 = e^{i(\pi/2)} = i$.

The quadratic character of 2 will now be derived from the relation

$$(\zeta + \zeta^{-1})^2 = \zeta^2 + 2 + \zeta^{-2} = 2$$

Let $\tau = \zeta + \zeta^{-1}$ and notice that ζ and τ are algebraic integers. We may thus work with congruences in the ring of algebraic integers.

Let p be an odd prime in $\mathbb{Z}$ and notice that

$$\tau^{p-1} = (\tau^2)^{(p-1)/2} = 2^{(p-1)/2} \equiv (2/p)\,(p)$$

It follows that $\tau^p \equiv (2/p)\tau\,(p)$. By Proposition 6.1.6, $\tau^p = (\zeta + \zeta^{-1})^p \equiv \zeta^p + \zeta^{-p}\,(p)$.

Remembering that $\zeta^8 = 1$ we have $\zeta^p + \zeta^{-p} = \zeta + \zeta^{-1}$ for $p \equiv \pm 1\,(8)$ and $\zeta^p + \zeta^{-p} = \zeta^3 + \zeta^{-3}$ for $p \equiv \pm 3\,(8)$. The result in the latter case may be simplified by observing that $\zeta^4 = -1$ implies that $\zeta^3 = -\zeta^{-1}$. Thus $\zeta^p + \zeta^{-p} = -(\zeta + \zeta^{-1})$ if $p \equiv \pm 3\,(8)$. Summarizing,

$$\zeta^p + \zeta^{-p} = \begin{cases} \tau & \text{if } p \equiv \pm 1\,(8) \\ -\tau & \text{if } p \equiv \pm 3\,(8) \end{cases}$$

Substituting this result into the relation $\tau^p \equiv (2/p)\tau\,(p)$ yields

$$(-1)^\varepsilon \tau \equiv (2/p)\tau\,(p), \qquad \text{where } \varepsilon \equiv \frac{p^2 - 1}{8}\,(2)$$

Multiply both sides of the congruence by τ. Then

$$(-1)^\varepsilon 2 \equiv (2/p)2\,(p)$$

implying that

$$(-1)^{\varepsilon} \equiv (2/p)\ (p)$$

This last congruence implies that $(2/p) = (-1)^{\varepsilon}$, which is the desired result.

Euler (1707–1783), in an early paper, proved that 2 is a quadratic residue modulo primes $p \equiv 1\ (8)$. His method contains the key idea of the above proof.

Euler assumed that $U(\mathbb{Z}/p\mathbb{Z})$ is a cyclic group. Gauss was the first to give a rigorous proof of this fact (see Theorem 1 of Chapter 4). Let λ be a generator of $U(\mathbb{Z}/p\mathbb{Z})$ and set $\gamma = \lambda^{(p-1)/8}$. Then γ has order 8, so that $\gamma^4 = -\bar{1}$ and $\gamma^2 + \gamma^{-2} = \bar{0}$. Therefore, $(\gamma + \gamma^{-1})^2 = \gamma^2 + \bar{2} + \gamma^{-2} = \bar{2}$. This shows that $\bar{2}$ is a square in $U(\mathbb{Z}/p\mathbb{Z})$, which is equivalent to 2 being a quadratic residue modulo p.

If $p \not\equiv 1\ (8)$, this proof cannot get started. However, the theory of finite fields enables us to carry through to a complete proof of quadratic reciprocity using Euler's idea. We shall develop the theory of finite fields in Chapter 7.

3 QUADRATIC GAUSS SUMS

Given the relation $(\zeta + \zeta^{-1})^2 = 2$ of Section 2, one might ask if there is a similar relation when 2 is replaced by an odd prime p. The answer is yes, and, moreover, the full law of quadratic reciprocity follows from this new relation by using the method of Section 2.

Throughout this section ζ will denote $e^{2\pi i/p}$, a primitive pth root of unity.

Lemma 1

$\sum_{t=0}^{p-1} \zeta^{at}$ is equal to p if $a \equiv 0\ (p)$. Otherwise it is zero.

PROOF

If $a \equiv 0\ (p)$, then $\zeta^a = 1$, and so $\sum_{t=0}^{p-1} \zeta^{at} = p$. If $a \not\equiv 0\ (p)$, then $\zeta^a \neq 1$ and $\sum_{t=0}^{p-1} \zeta^{at} = (\zeta^{ap} - 1)/(\zeta^a - 1) = 0$.

Corollary

$p^{-1} \sum_{t=0}^{p-1} \zeta^{t(x-y)} = \delta(x, y)$, where $\delta(x, y) = 1$ if $x \equiv y\ (p)$ and $\delta(x, y) = 0$ if $x \not\equiv y\ (p)$.

PROOF

The proof is immediate from Lemma 1.

All summations for the remainder of this section will be over the range zero to $p-1$. It will simplify notation to avoid writing out this fact each time.

Lemma 2

$\sum_t (t/p) = 0$, *where* (t/p) *is the Legendre symbol.*

PROOF

By definition $(0/p) = 0$. Of the remaining $p-1$ terms in the summation, half are $+1$ and half are -1, since by Corollary 1 to Proposition 5.1.2, there are as many quadratic residues as quadratic nonresidues mod p.

We are now in a position to introduce the notion of Gauss sum.

Definition

$g_a = \sum_t (t/p)\zeta^{at}$ is called a *quadratic Gauss sum.*

Proposition 6.3.1

$g_a = (a/p)g_1$.

PROOF

If $a \equiv 0\,(p)$, then $\zeta^{at} = 1$ for all t, and $g_a = \sum(t/p) = 0$ by Lemma 2. This gives the result in the case that $a \equiv 0\,(p)$.
Now suppose that $a \not\equiv 0\,(p)$. Then

$$(a/p)g_a = \sum_t (at/p)\zeta^{at} = \sum_x (x/p)\zeta^x = g_1$$

We have used the fact that at runs over a complete residue system mod p when t does and that (x/p) and ζ^x depend only on the residue class of x modulo p.

Since $(a/p)^2 = 1$ when $a \not\equiv 0\,(p)$ our result follows by multiplying the equation $(a/p)g_a = g_1$ on both sides by (a/p).

From now on we shall denote g_1 by g. It follows from Proposition 6.3.1 that $g_a^2 = g^2$ if $a \not\equiv 0\,(p)$. We shall now deduce this common value.

Proposition 6.3.2

$g^2 = (-1)^{(p-1)/2}p$.

PROOF

The idea of the proof is to evaluate the sum $\sum_a g_a g_{-a}$ in two ways.

If $a \not\equiv 0\ (p)$, then $g_a g_{-a} = (a/p)(-a/p)g^2 = (-1/p)g^2$. It follows that

$$\sum_a g_a g_{-a} = (-1/p)(p-1)g^2$$

Now, notice that

$$g_a g_{-a} = \sum_x \sum_y (x/p)(y/p)\zeta^{a(x-y)}$$

Summing both sides over a and using the corollary to Lemma 1 yields

$$\sum_a g_a g_{-a} = \sum_x \sum_y (x/p)(y/p)\delta(x, y)p = (p-1)p$$

Putting these results together we obtain $(-1/p)(p-1)g^2 = (p-1)p$. Therefore, $g^2 = (-1/p)p$.

Let $p^* = (-1)^{(p-1)/2}p$. The equation $g^2 = p^*$ is the desired analog of the equation $\tau^2 = 2$. Let $q \neq p$ be another odd prime. We proceed to prove the law of quadratic reciprocity by working with congruences mod q in the ring of algebraic integers:

$$g^{q-1} = (g^2)^{(q-1)/2} = p^{*(q-1)/2} \equiv (p^*/q)\ (q)$$

Thus

$$g^q \equiv (p^*/q)g\ (q)$$

Using Proposition 6.1.6 we see that

$$g^q = \left(\sum (t/p)\zeta^t\right)^q \equiv \sum (t/p)^q \zeta^{qt} \equiv g_q\ (q)$$

It follows that $g^q \equiv g_q \equiv (q/p)g\ (q)$ and so

$$(q/p)g \equiv (p^*/q)g\ (q)$$

Multiply both sides by g, and use $g^2 = p^*$:

$$(q/p)p^* \equiv (p^*/q)p^*\ (q)$$

which implies that

$$(q/p) \equiv (p^*/q)\ (q)$$

and finally

$$(q/p) = (p^*/q)$$

To see that this result is what we want simply notice that

$$(p^*/q) = (-1/q)^{(p-1)/2}(p/q) = (-1)^{((q-1)/2)((p-1)/2)}(p/q)$$

The notion of quadratic Gauss sum that we have used can be considerably generalized. We shall present some of these generalizations after developing the theory of finite fields. Cubic Gauss sums will be used to prove the law of cubic reciprocity.

Notes

Excellent introductions to the theory of algebraic numbers are found in Samuel [68], H. Mann [58], Pisot [62], Borevich and Shafarevich [9], and Hecke [44].

Gauss worked very hard to determine the Gauss sum g, which is defined in Section 3. Since, as we have seen, $g^2 = p^*$, what is at stake is the sign of the Gauss sum. After working diligently on this problem for over a year and getting nowhere, the answer suddenly came to Gauss, "... the way lightening strikes," as he described it in a letter to H. W. M. Olbers. The answer is $g = \sqrt{p}$ if $p \equiv 1$ (4) and $g = i\sqrt{p}$ if $p \equiv 3$ (4). There are numerous proofs of this fact now in existence. For a discussion, see Chowla [18]. The most recent proof is given in a short note by W. Waterhouse [79] in which he simplifies an old proof due to I. Schur.

For the history of the algebraic theory of numbers, we mention the interesting paper of L. Dickson, Fermat's Last Theorem and the Origin and Nature of Algebraic Numbers (*Annals of Math.*, **20**, 1919, 155–171).

Exercises

1 Show that $\sqrt{2} + \sqrt{3}$ is an algebraic integer.

2 Let α be an algebraic number. Show that there is an integer n such that $n\alpha$ is an algebraic integer.

3 If α and β are algebraic integers, prove that any solution to $x^2 + \alpha x + \beta = 0$ is an algebraic integer. Generalize this result.

4 A polynomial $f(x) \in \mathbb{Z}[x]$ is said to be primitive if the greatest common divisor of its coefficients is 1. Prove that the product of primitive polynomials is again primitive. [*Hint:* Let $f(x) = a_0x^n + a_1x^{n-1} + \cdots + a_n$ and $g(x) = b_0x^m + b_1x^{m-1} + \cdots + b_m$ be primitive. If p is a prime, let a_i and b_j be the coefficients with the smallest subscripts such that $p \nmid a_i$ and $p \nmid b_j$. Show that the coefficient of x^{i+j} in $f(x)g(x)$ is not divisible by p.] This is one of the many results known as Gauss's lemma.

5 Let α be an algebraic integer and $f(x) \in \mathbb{Q}[x]$ be the monic polynomial of least degree such that $f(\alpha) = 0$. Use Exercise 4 to show that $f(x) \in \mathbb{Z}[x]$.

6 Let $x^2 + mx + n \in \mathbb{Z}[x]$ be irreducible and α be a root. Show that $\mathbb{Q}[\alpha] = \{r + s\alpha \mid r, s \in \mathbb{Q}\}$ is a ring (in fact, it is a field). Let $m^2 - 4n = D_0^2D$, where D is square-free. Show that $\mathbb{Q}[\alpha] = \mathbb{Q}[\sqrt{D}]$.

7 (continuation) If $D \equiv 2, 3$ (4), show that all the algebraic integers in $\mathbb{Q}[\sqrt{D}]$ have the form $a + b\sqrt{D}$, where $a, b \in \mathbb{Z}$. If $D \equiv 1$ (4), show that all the algebraic integers in $\mathbb{Q}[\sqrt{D}]$ have the form $a + b((-1 + \sqrt{D})/2)$, where $a, b \in \mathbb{Z}$. [*Hint:* Show that $r + s\sqrt{D}$ satisfies $x^2 - 2rx + (r^2 - Ds^2) = 0$. Thus by Exercise 5 $r + s\sqrt{D}$ is an algebraic integer iff $2r$ and $r^2 - Ds^2$ are in $\mathbb{Z}$.]

8 Let $\omega = e^{2\pi i/3}$. ω satisfies $x^3 - 1 = 0$. Show that $(2\omega + 1)^2 = -3$ and use this to determine $(-3/p)$ by the method of Section 2.

9 Verify Proposition 6.3.2 explicitly for $p = 3$ and $p = 5$; i.e., write out the Gauss sum longhand and square.

10 What is $\sum_{a=1}^{p-1} g_a$?

11 By evaluating $\sum_t (1 + (t/p))\zeta^t$ in two ways prove that $g = \sum_t \zeta^{t^2}$.

12 Write $\psi_a(t) = \zeta^{at}$. Show that

(a) $\overline{\psi_a(t)} = \psi_a(-t) = \psi_{-a}(t)$.

(b) $(1/p) \sum_a \psi_a(t - s) = \delta(t, s)$.

13 Let f be a function from $\mathbb{Z}$ to the complex numbers. Suppose that p is a prime and that $f(n + p) = f(n)$ for all $n \in \mathbb{Z}$. Let $\hat{f}(a) = p^{-1} \sum_t f(t)\psi_{-a}(t)$. Prove that $f(t) = \sum_a \hat{f}(a)\psi_a(t)$. This result is directly analogous to a result in the theory of Fourier series.

14 In Exercise 13 take f to be the Legendre symbol and show that $\hat{f}(a) = p^{-1}g_{-a}$.

15 Show that $|\sum_{t=m}^{n} (t/p)| < \sqrt{p} \log p$. The inequality holds for the sum over any range. This remarkable inequality is associated with the names of Polya and Vinogradov. [*Hint:* Use the relation $(t/p)g = g_t$ and sum. The inequality $\sin x \geq (2/\pi)x$ for any acute angle x will be useful.]

chapter seven/FINITE FIELDS

We have already met with examples of finite fields, namely, the fields $\mathbb{Z}/p\mathbb{Z}$, where p is a prime number. In this chapter we shall prove that there are many more finite fields and shall investigate their properties. This theory is beautiful and interesting in itself and, moreover, is a very useful tool in number-theoretic investigations. As an illustration of the latter point, we shall supply yet another proof of the law of quadratic reciprocity. Other applications will come later.

One more comment. Up to now the great majority of our proofs have used very few results from abstract algebra. Although nowhere in this book will we use very sophisticated results from algebra, from now on we shall assume that the reader has some familiarity with the material in a standard undergraduate course in the subject.

1 *BASIC PROPERTIES OF FINITE FIELDS*

In this section we shall discuss properties of finite fields without worrying about questions of existence. The construction of finite fields will be taken up in Section 2.

Let F be a finite field with q elements. The multiplicative group of F, F^*, has $q - 1$ elements. Thus every element $\alpha \in F^*$ satisfies the equation $x^{q-1} = 1$ (in this context 1 stands for the multiplicative identity of F and not the integer 1), and every element in F satisfies $x^q = x$.

Proposition 7.1.1

$$x^q - x = \prod_{\alpha \in F} (x - \alpha)$$

PROOF

Both polynomials are to be considered as elements of $F[x]$.

Every element $\alpha \in F$ is a root of $x^q - x$. Since F has q elements and since the degree of $x^q - x$ is q, the result follows.

Corollary 1

Let $F \subset K$, where K is a field. An element $\alpha \in K$ is in F iff $\alpha^q = \alpha$.

PROOF

$\alpha^q = \alpha$ iff α is a root of $x^q - x$. By Proposition 7.1.1, the roots of $x^q - x$ are precisely the elements of F.

Corollary 2

If $f(x)$ divides $x^q - x$, then $f(x)$ has d distinct roots, where d is the degree of $f(x)$.

PROOF

Let $f(x)g(x) = x^q - x$. $g(x)$ has degree $q - d$. If $f(x)$ has fewer than d distinct roots, then by Lemma 1 of Chapter 4, $f(x)g(x)$ would have fewer than $d + (q - d) = q$ distinct roots, which is not the case.

Theorem 1

The multiplicative group of a finite field is cyclic.

PROOF

This theorem is a generalization of Theorem 1 in Chapter 4. The proof is almost identical.

If $d \mid q - 1$, then $x^d - 1$ divides $x^{q-1} - 1$ and it follows from Corollary 2 that $x^d - 1$ has d distinct roots. Thus the subgroup of F^* consisting of elements satisfying $x^d = 1$ has order d.

Let $\psi(d)$ be the number of elements in F^* of order d. Then $\sum_{c|d} \psi(c) = d$. By the Möbius inversion formula

$$\psi(d) = \sum_{c|d} \mu(c)\frac{d}{c} = \phi(d)$$

In particular, $\psi(q - 1) = \phi(q - 1) > 1$, unless we are in the trivial case $q = 2$. This concludes the proof.

The fact that F^* is cyclic when F is finite allows us to give the following partial generalization of Proposition 4.2.1.

Proposition 7.1.2

Let $\alpha \in F^*$. Then $x^n = \alpha$ has solutions iff $\alpha^{(q-1)/d} = 1$, where $d = (n, q - 1)$. If there are solutions, then there are exactly d solutions.

PROOF

Let γ be a generator of F^* and set $\alpha = \gamma^a$ and $x = \gamma^y$. Then $x^n = \alpha$ is equivalent to the congruence $ny \equiv a\ (q - 1)$. The result now follows by applying Proposition 3.3.1.

It is worthwhile to examine what happens in the extreme cases $n \mid q - 1$ and $(n, q - 1) = 1$.

If $n \mid q - 1$, then there are exactly $(q - 1)/n$ elements of F^* that are nth powers, and if α is an nth power, then $x^n = \alpha$ has n solutions.

If $(n, q - 1) = 1$, then every element is an nth power in a unique way; i.e., for $\alpha \in F^*$, $x^n = \alpha$ has one and only one solution.

We have investigated the structure of F^*. Now we turn our attention to the additive group of F.

Lemma 1

Let F be a finite field. The integer multiples of the identity form a subfield of F isomorphic to $\mathbb{Z}/p\mathbb{Z}$ for some prime number p.

PROOF

To avoid confusion, let us temporarily call e the identity of F^* instead of 1. Map $\mathbb{Z}$ to F by taking n to ne. This is easily seen to be a ring homomorphism. The image is a finite subring of F, and so in particular it is an integral domain. The kernel is a nonzero prime ideal. Therefore, the image is isomorphic to $\mathbb{Z}/p\mathbb{Z}$ for some prime p.

We shall identify $\mathbb{Z}/p\mathbb{Z}$ with its image in F and think of F as a finite dimensional vector space over $\mathbb{Z}/p\mathbb{Z}$. Let n denote that dimension and let $\omega_1, \omega_2, \ldots, \omega_n$ be a basis. Then every element $\omega \in F$ can be expressed uniquely in the form $a_1\omega_1 + a_2\omega_2 + \cdots + a_n\omega_n$, where $a_i \in \mathbb{Z}/p\mathbb{Z}$. It follows that F has p^n elements. We have proved

Proposition 7.1.3

The number of elements in a finite field is a power of a prime.

If e is the identity of the finite field F, let p be the smallest integer such that $pe = 0$. We have seen that p must be a prime number. It is called the characteristic of F. For $\alpha \in F$ we have $p\alpha = p(e\alpha) = (pe)\alpha = 0 \cdot \alpha = 0$. This observation leads to the following very useful proposition.

Proposition 7.1.4

If F has characteristic p, then $(\alpha + \beta)^{p^d} = \alpha^{p^d} + \beta^{p^d}$ for all $\alpha, \beta \in F$ and all positive integers d.

PROOF

The proof is by induction on d. For $d = 1$, we have

$$(\alpha + \beta)^p = \alpha^p + \sum_{k=1}^{p-1} \binom{p}{k} \alpha^{p-k}\beta^k + \beta^p = \alpha^p + \beta^p$$

All the intermediate terms vanish because $p \,\Big|\, \binom{p}{k}$ for $1 \leq k \leq p - 1$ by Lemma 2 of Chapter 4.

To pass from d to $d + 1$ just raise both sides of $(\alpha + \beta)^{p^d} = \alpha^{p^d} + \beta^{p^d}$ to the pth power.

Suppose that F is a finite field of dimension n over $\mathbb{Z}/p\mathbb{Z}$. We want to find out which fields E lie between $\mathbb{Z}/p\mathbb{Z}$ and F. If d is the dimension of E over $\mathbb{Z}/p\mathbb{Z}$, then it follows by general field theory that $d \mid n$. We shall give another proof below. It turns out that there is one and only one intermediate field corresponding to every divisor d of n.

Lemma 2
Let F be a field. Then $x^l - 1$ divides $x^m - 1$ in $F[x]$ iff l divides m.

PROOF
Let $m = ql + r$, where $0 \leq r < l$. Then we have

$$\frac{x^m - 1}{x^l - 1} = x^r \frac{x^{ql} - 1}{x^l - 1} + \frac{x^r - 1}{x^l - 1}$$

Since $(x^{ql} - 1)/(x^l - 1) = (x^l)^{q-1} + (x^l)^{q-2} + \cdots + x^l + 1$, the right-hand side of the above equation is a polynomial iff $(x^r - 1)/(x^l - 1)$ is a polynomial. This is easily seen to be the case iff $r = 0$. The result follows.

Lemma 3
If a is a positive integer, then $a^l - 1$ divides $a^m - 1$ iff l divides m.

PROOF
The proof is analogous to that of Lemma 2 with the number a playing the role of x. We leave the details to the reader.

Proposition 7.1.5
Let F be a finite field of dimension n over $\mathbb{Z}/p\mathbb{Z}$. The subfields of F are in one-to-one correspondence with the divisors of n.

PROOF
Suppose that E is a subfield of F and let d be its dimension over $\mathbb{Z}/p\mathbb{Z}$. We shall show that $d \mid n$.

Since E^* has $p^d - 1$ elements all satisfying $x^{p^d - 1} - 1$, we have that $x^{p^d - 1} - 1$ divides $x^{p^n - 1} - 1$. By Lemma 2 $p^d - 1$ divides $p^n - 1$ and consequently by Lemma 3 d divides n.

Now suppose that $d \mid n$. Let $E = \{\alpha \in F \mid \alpha^{p^d} = \alpha\}$. We claim that E is a field. For if $\alpha, \beta \in E$, then

(a) $(\alpha + \beta)^{p^d} = \alpha^{p^d} + \beta^{p^d} = \alpha + \beta$.

(b) $(\alpha\beta)^{p^d} = \alpha^{p^d}\beta^{p^d} = \alpha\beta$.

(c) $(\alpha^{-1})^{p^d} = (\alpha^{p^d})^{-1} = \alpha^{-1}$ for $\alpha \neq 0$.

In step (a) we made use of Proposition 7.1.4.

Now E is the set of solutions to $x^{p^d} - x$. Since $d \mid n$, we have $p^d - 1 \mid p^n - 1$ and $x^{p^d - 1} - 1 \mid x^{p^n - 1} - 1$ by Lemmas 2 and 3. Thus $x^{p^d} - x$ divides $x^{p^n} - x$, and by Corollary 2 to Proposition 7.1.1, it follows that E has p^d elements and so has dimension d over $\mathbb{Z}/p\mathbb{Z}$.

Finally, if E' is another subfield of F of dimension d over $\mathbb{Z}/p\mathbb{Z}$, then the elements of E' must satisfy $x^{p^d} - x = 0$; i.e., E' must coincide with E.

Let F_q denote a finite field with q elements. To illustrate Proposition 7.1.5, consider F_{4096} (we shall show in Section 2 the existence of such a field). Since $4096 = 2^{12}$ we have the following lattice diagram:

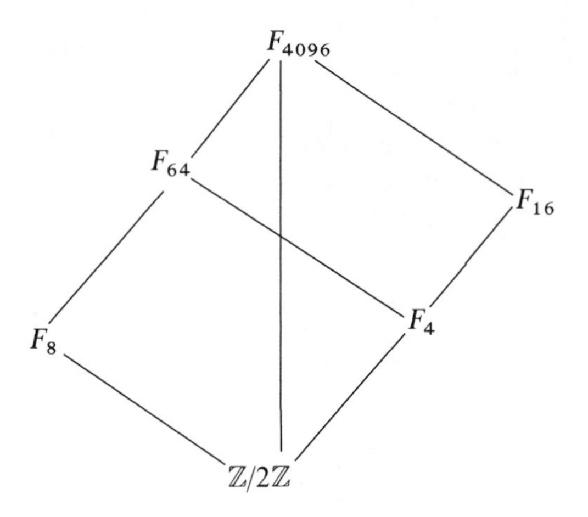

2 THE EXISTENCE OF FINITE FIELDS

In Section 1 we proved that the number of elements in a finite field has the form p^n, where p is a prime. We shall now show that given a number p^n there exists a finite field with p^n elements. To do this we shall need some results from the theory of fields that connect our problem with the existence of irreducible polynomials. Then we shall prove a theorem that goes back to Gauss (again!) that shows that $\mathbb{Z}/p\mathbb{Z}[x]$ contains irreducible polynomials of every degree.

Let k be an arbitrary field and $f(x)$ be an irreducible polynomial in $k[x]$. We then have

Proposition 7.2.1
There exists a field K containing k and an element $\alpha \in K$ such that $f(\alpha) = 0$.

PROOF
We proved in Chapter 1 that $k[x]$ is a principal ideal domain. It follows that $(f(x))$ is a maximal ideal and thus $k[x]/(f(x))$ is a field. Let $K' = k[x]/(f(x))$ and let ϕ be the homomorphism that maps $k[x]$ onto K' by taking an element to its coset modulo $(f(x))$. We have the diagram

$$\begin{array}{ccc} k[x] & \xrightarrow{\phi} & K' \\ | & & | \\ k & \xrightarrow{\phi} & \phi(k) \end{array}$$

$\phi(k)$ is a subfield of K'. We claim that it is isomorphic to k. It is enough to show that ϕ is one to one. Let $a \in k$. If $\phi(a) = 0$, then $a \in (f(x))$. If $a \neq 0$, it is a unit and cannot be an element of a proper ideal. Thus $a = 0$, as was to be shown.

Since ϕ is an isomorphism of k we may identify k with $\phi(k)$. When this is done we relabel K' as K.

Let α be the coset of x in K. Then $0 = \phi(f(x)) = f(\phi(x)) = f(\alpha)$; i.e., α is a root of $f(x)$ in K.

We denote the field K constructed in the proposition by $k(\alpha)$. The following proposition about $k(\alpha)$ will be useful.

Proposition 7.2.2
The elements $1, \alpha, \alpha^2, \ldots, \alpha^{n-1}$ are a vector space basis for $k(\alpha)$ over k, where n is the degree of $f(x)$.

PROOF
Let $f(x) = a_0x^n + a_1x^{n-1} + \cdots + a_{n-1}x + a_n$, where the $a_i \in k$ and $a_0 \neq 0$.

By the construction in Proposition 7.2.1 every element in K is a polynomial in α with coefficients in k. We shall have shown that $1, \alpha, \ldots, \alpha^{n-1}$ span $k(\alpha)$ over k if we can show that α^m for $m > n - 1$ is a k linear combination of $1, \alpha, \ldots, \alpha^{n-1}$.

Consider α^n. Since $f(\alpha) = 0$ it follows that

$$\alpha^n = -a_0^{-1}(a_1\alpha^{n-1} + \cdots + a_{n-1}\alpha + a_n)$$

which proves what we want for α^n. Now,

$$\alpha^{n+k} = -a_0^{-1}(a_1\alpha^{n+k-1} + \cdots + a_{n-1}\alpha^{k+1} + a_n\alpha^k)$$

so the result follows by induction.

It remains to prove that $1, \alpha, \ldots, \alpha^{n-1}$ are linearly independent over k. Suppose that $b_1\alpha^{n-1} + b_2\alpha^{n-2} + \cdots + b_{n-1}\alpha + b_n = 0$. Let $g(x) = b_1x^{n-1} + \cdots + b_{n-1}x + b_n$. If $g(x)$ is not the zero polynomial, it is relatively prime to $f(x)$ since $f(x)$ is irreducible and $g(x)$ has smaller degree than $f(x)$. Thus there exist polynomials $h(x)$ and $l(x)$ such that $h(x)f(x) + l(x)g(x) = 1$. Substituting $x = \alpha$ yields $0 = 1$, a contradiction. Thus $b_1 = b_2 = \cdots = b_n = 0$.

Corollary

The dimension of $k(\alpha)$ over k is n, the degree of $f(x)$.

To turn the matter around, the corollary shows that if we want to find a field extension K of k of degree n, then it is enough to produce an irreducible polynomial $f(x) \in k[x]$ of degree n.

In $\mathbb{Z}/p\mathbb{Z}[x]$ there are finitely many polynomials of a given degree. Let $F_d(x)$ be the product of the monic irreducible polynomials in $\mathbb{Z}/p\mathbb{Z}[x]$ of degree d.

Theorem 1

$$x^{p^n} - x = \prod_{d|n} F_d(x)$$

PROOF

First notice that if $f(x)$ divides $x^{p^n} - x$, then $f(x)^2$ does not divide $x^{p^n} - x$. This follows since if $x^{p^n} - x = f(x)^2g(x)$ we obtain

$$-1 = 2f(x)f'(x)g(x) + f(x)^2g'(x)$$

by formal differentiation. This is impossible since it implies that $f(x)$ divides 1.

It remains to prove that if $f(x)$ is a monic irreducible polynomial of degree d, then $f(x) \mid x^{p^n} - x$ iff $d \mid n$.

Consider $K = \mathbb{Z}/p\mathbb{Z}(\alpha)$, where α is a root of $f(x)$, as in Proposition 7.2.2. It has dimension d over $\mathbb{Z}/p\mathbb{Z}$ and thus p^d elements. The elements of K satisfy $x^{p^d} - x = 0$.

Assume that $x^{p^n} - x = f(x)g(x)$. Then $\alpha^{p^n} = \alpha$. If $b_1\alpha^{d-1} + b_2\alpha^{d-2} + \cdots + b_d$ is an arbitrary element of K, then

$$(b_1\alpha^{d-1} + \cdots + b_d)^{p^n} = b_1(\alpha^{p^n})^{d-1} + \cdots + b_d = b_1\alpha^{d-1} + \cdots + b_d$$

Hence the elements of K satisfy $x^{p^n} - x = 0$. It follows that $x^{p^d} - x$ divides $x^{p^n} - x$, and by Lemmas 2 and 3 of Section 1 d divides n.

Assume now that $d \mid n$. Since $\alpha^{p^d} = \alpha$ and $f(x)$ is the monic irreducible polynomial for α, we have $f(x) \mid x^{p^d} - x$. Since $d \mid n$ we have $x^{p^d} - x \mid x^{p^n} - x$ again by Lemmas 2 and 3 of Section 1. Thus $f(x) \mid x^{p^n} - x$.

Let N_d be the number of monic irreducible polynomials of degree d in $\mathbb{Z}/p\mathbb{Z}[x]$. Equating the degrees on both sides of the identity in the theorem yields

Corollary 1
$p^n = \sum_{d \mid n} dN_d$.

Corollary 2
$N_n = n^{-1} \sum_{d \mid n} \mu(n/d)p^d$.

PROOF
Apply the Möbius inversion formula (Theorem 1 of Chapter 2) to the equation in Corollary 1.

Corollary 3
For each integer $n \geq 1$, *there exists an irreducible polynomial of degree* n *in* $\mathbb{Z}/p\mathbb{Z}[x]$.

PROOF
$N_n = n^{-1}(p^n - \cdots + p\mu(n))$ by Corollary 2. The term in parentheses cannot be zero since it is the sum of distinct powers of p with coefficients 1 and -1.

Summarizing, we have

Theorem 2
Let $n \geq 1$ *be an integer and* p *be a prime. Then there exists a finite field with* p^n *elements.*

3 AN APPLICATION TO QUADRATIC RESIDUES

In Chapter 6 we proved the law of quadratic reciprocity using Gauss sums and the elements of the theory of algebraic numbers. We shall now give an exceptionally short proof along the same lines using the theory of finite fields.

Let p and q be distinct odd primes. Since $(p, q) = 1$ there is an integer n (for example, $p - 1$) such that $q^n \equiv 1\ (p)$. Let F be a finite field of dimension n over $\mathbb{Z}/q\mathbb{Z}$. Then F^* is cyclic of order $q^n - 1$. Let γ be a generator of F^* and set $\lambda = \gamma^{(q^n-1)/p}$. Then λ has order p. Define $\tau_a = \sum_{t=0}^{p-1} (t/p)\lambda^{at}$, where $a \in \mathbb{Z}$. The element τ_a of F is an analog of the quadratic Gauss sums introduced in Chapter 6. Set $\tau_1 = \tau$. Then the proofs of Propositions 6.3.1 and 6.3.2 can be used to show that

1. $\tau_a = (a/p)\tau$.
2. $\tau^2 = (-1)^{(p-1)/2}\bar{p}$.

In relation 2, $\bar{p}$ is the coset of p in $\mathbb{Z}/q\mathbb{Z}$. Let $p^* = (-1)^{(p-1)/2}p$. Then relation 2 can be written as $\tau^2 = \overline{p^*}$. This relation implies that $(p^*/q) = 1$ iff $\tau \in \mathbb{Z}/q\mathbb{Z}$. By Corollary 1 to Proposition 7.1.1, this is true iff $\tau^q = \tau$. Now,

$$\tau^q = \left(\sum_t (t/p)\lambda^t\right)^q = \sum_t (t/p)\lambda^{qt} = \tau_q$$

By relation 1 we have $\tau_q = (q/p)\tau$. Thus $\tau^q = \tau$ iff $(q/p) = 1$. We have proved that

$$(p^*/q) = 1 \quad \text{iff} \quad (q/p) = 1$$

This is the law of quadratic reciprocity.

A proof that $(2/q) = (-1)^{(q^2-1)/8}$ can be given using the same technique. In Chapter 6 we gave Euler's proof that $(2/q) = 1$ if $q \equiv 1\ (8)$. If $q \not\equiv 1\ (8)$, it is nevertheless true that $q^2 \equiv 1\ (8)$. In this case one can carry through the proof working in a finite field F of dimension 2 over $\mathbb{Z}/q\mathbb{Z}$. We leave the details to the reader.

Notes

The first systematic account of the theory of finite fields is found in Dickson [25], although E. Galois had axiomatically developed a number of their properties much earlier in his note "Sur la théorie des nombres" [33]. As the existence of a finite field with p^n elements is equivalent to the existence of an irreducible polynomial of degree n in the ring $F[x]$ we must include Gauss once again as a founder. In his paper "Die Lehre von den Reste" he derives the formula we have given for the number of irreducibles of degree n (see [34]).

The use of finite fields to give a proof of quadratic reciprocity has been observed by a number of mathematicians, e.g., Hausner [43] and Holzer [45, pp. 76–78].

Our treatment of finite fields throughout this book is much more elementary than is usual in modern times. Most treatments first develop the full Galois theory of fields and apply the general results of that theory to the special case of finite fields. This is done in A. Albert's compact book [1]. The advantage of Albert's book for those readers already familiar with the theory of fields is that he discusses finite fields extensively in his last chapter and provides a very long bibliography on the subject. Many interesting references are provided.

Exercises

1 Use the method of Theorem 1 to show that a finite subgroup of the multiplicative group of a field is cyclic.

2 Let R and C be the real and complex numbers, respectively. Find the finite subgroups of R^* and C^* and show directly that they are cyclic.

3 Let F be a field with q elements and suppose that $q \equiv 1\ (n)$. Show that for $\alpha \in F^*$ the equation $x^n = \alpha$ has either no solutions or n solutions.

4 (continuation) Show that the set of $\alpha \in F^*$ such that $x^n = \alpha$ is solvable is a subgroup with $(q-1)/n$ elements.

5 (continuation) Let K be a field containing F such that $[K:F] = n$. For all $\alpha \in F^*$ show that the equation $x^n = \alpha$ has n solutions in K. [*Hint:* Show that $q^n - 1$ is divisible by $n(q-1)$ and use the fact that $\alpha^{q-1} = 1$.]

6 Let $K \supset F$ be finite fields with $[K:F] = 3$. Show that if $\alpha \in F$ is not a square in F, it is not a square in K.

7 Generalize Exercise 6 by showing that if α is not a square in F, it is not a square in any extension of odd degree and is a square in every extension of even degree.

8 In a field with 2^n elements what is the subgroup of squares?

9 If $K \supset F$ are finite fields, $|F| = q, \alpha \in F, q \equiv 1\ (n)$, and $x^n = \alpha$ is not solvable in F, show that $x^n = \alpha$ is not solvable in K if $(n, [K:F]) = 1$.

10 Let $K \supset F$ be finite fields and $[K:F] = 2$. For $\beta \in K$ show that $\beta^{1+q} \in F$ and moreover that every element in F is of the form β^{1+q} for some $\beta \in K$.

11 With the situation being that of Exercise 10 suppose that $\alpha \in F$ has order $q-1$. Show that there is a $\beta \in K$ with order $q^2 - 1$ such that $\beta^{1+q} = \alpha$.

12 Use Proposition 7.2.1 to show that given a field k and a polynomial $f(x) \in k[x]$ there is a field $K \supset k$ such that $[K:k]$ is finite and $f(x) = (x-\alpha_1)(x-\alpha_2)\cdots(x-\alpha_n)$ in $K[x]$.

13 Apply Exercise 12 to $k = \mathbb{Z}/p\mathbb{Z}$ and $f(x) = x^{p^n} - x$ to obtain another proof of Theorem 2.

14 Let F be a field with q elements and n a positive integer. Show that there exist irreducible polynomials in $F[x]$ of degree n.

15 Let $x^n - 1 \in F[x]$, where F is a finite field with q elements. Suppose that $(q, n) = 1$. Show that $x^n - 1$ splits into linear factors in some extension field and that the least degree of such a field is the smallest integer f such that $q^f \equiv 1\ (n)$.

16 Calculate the monic irreducible polynomials of degree 4 in $\mathbb{Z}/2\mathbb{Z}[x]$.

17 Let q and p be distinct odd primes. Show that the number of monic irreducibles of degree q in $\mathbb{Z}/p\mathbb{Z}[x]$ is $q^{-1}(p^q - p)$.

18 Let p be a prime with $p \equiv 3\ (4)$. Show that the residue classes modulo p in $\mathbb{Z}[i]$ form a field with p^2 elements.

19 Let F be a finite field with q elements. If $f(x) \in F[x]$ has degree t, put $|f| = q^t$. Verify the formal identity $\sum_f |f|^{-s} = (1 - q^{1-s})^{-1}$. The sum is over all monic polynomials.

20 With the notation of Exercise 19 let $d(f)$ be the number of monic divisors of f and $\sigma(f) = \sum_{g|f} |g|$, where the sum is over the monic divisors of f. Verify the following identities:
(a) $\sum_f d(f)|f|^{-s} = (1 - q^{1-s})^{-2}$.
(b) $\sum_f \sigma(f)|f|^{-s} = (1 - q^{1-s})^{-1}(1 - q^{2-s})^{-1}$.

21 Let F be a field with $q = p^n$ elements. For $\alpha \in F$ set $f(x) = (x - \alpha)(x - \alpha^p) \times (x - \alpha^{p^2}) \cdots (x - \alpha^{p^{n-1}})$. Show that $f(x) \in \mathbb{Z}/p\mathbb{Z}[x]$. In particular, $\alpha + \alpha^p + \cdots + \alpha^{p^{n-1}}$ and $\alpha\alpha^p\alpha^{p^2} \cdots \alpha^{p^{n-1}}$ are in $\mathbb{Z}/p\mathbb{Z}$.

22 (continuation) Set $\operatorname{tr}(\alpha) = \alpha + \alpha^p + \cdots + \alpha^{p^{n-1}}$. Prove that
(a) $\operatorname{tr}(\alpha) + \operatorname{tr}(\beta) = \operatorname{tr}(\alpha + \beta)$.
(b) $\operatorname{tr}(a\alpha) = a \operatorname{tr}(\alpha)$ for $a \in \mathbb{Z}/p\mathbb{Z}$.
(c) There is an $\alpha \in F$ such that $\operatorname{tr}(\alpha) \neq 0$.

23 (continuation) For $\alpha \in F$ consider the polynomial $x^p - x - \alpha \in F[x]$. Show that this polynomial is either irreducible or the product of linear factors. Prove that the latter alternative holds iff $\operatorname{tr}(\alpha) = 0$.

24 Suppose that $f(x) \in \mathbb{Z}/p\mathbb{Z}[x]$ has the property that $f(x + y) = f(x) + f(y) \in \mathbb{Z}/p\mathbb{Z}[x, y]$. Show that $f(x)$ must be of the form $a_0x + a_1x^p + a_2x^{p^2} + \cdots + a_mx^{p^m}$.

chapter eight/GAUSS AND JACOBI SUMS

In Chapter 6 we introduced the notion of a quadratic Gauss sum. In this chapter a more general notion of Gauss sum will be introduced. These sums have many applications. They will be used in Chapter 9 as a tool in the proof of the law of cubic reciprocity. Here we shall consider the problem of counting the number of solutions of equations with coefficients in a finite field. In this connection, the notion of a Jacobi sum arises in a natural way. Jacobi sums are interesting in their own right, and we shall develop some of their properties.

To keep matters as simple as possible, we shall confine our attention to the finite field $\mathbb{Z}/p\mathbb{Z} = F_p$ *and come back later to the question of associating Gauss sums with an arbitrary finite field.*

I MULTIPLICATIVE CHARACTERS

A multiplicative character on F_p is a map χ from F_p^* to the nonzero complex numbers that satisfies

$$\chi(ab) = \chi(a)\chi(b) \quad \text{for all} \quad a, b \in F_p^*$$

The Legendre symbol, (a/p), is an example of such a character if it is regarded as a function of the coset of a modulo p.

Another example is the trivial multiplicative character defined by the relation $\varepsilon(a) = 1$ for all $a \in F_p^*$.

It is often useful to extend the domain of definition of a multiplicative character to all of F_p. If $\chi \neq \varepsilon$, we do this by defining $\chi(0) = 0$. For ε we define $\varepsilon(0) = 1$. The usefulness of these definitions will soon become apparent.

Proposition 8.1.1

Let χ *be a multiplicative character and* $a \in F_p^*$. *Then*

(a) $\chi(1) = 1$.

(b) $\chi(a)$ *is a* $(p-1)$st *root of unity.*

(c) $\chi(a^{-1}) = \chi(a)^{-1} = \overline{\chi(a)}$.

[In part (a) the 1 on the left-hand side is the unit of F_p, whereas the 1 on the right-hand side is the complex number 1. The bar in part (c) is complex conjugation.]

PROOF

$\chi(1) = \chi(1 \cdot 1) = \chi(1)\chi(1)$. Thus $\chi(1) = 1$, since $\chi(1) \neq 0$.

To prove part (b), notice that $a^{p-1} = 1$ implies that $1 = \chi(1) = \chi(a^{p-1}) = \chi(a)^{p-1}$.

To prove part (c), notice that $1 = \chi(1) = \chi(a^{-1}a) = \chi(a^{-1})\chi(a)$. This shows that $\chi(a^{-1}) = \chi(a)^{-1}$. The fact that $\chi(a)^{-1} = \overline{\chi(a)}$ follows from the fact that $\chi(a)$ is a complex number of absolute value 1 by part (b).

Proposition 8.1.2

Let χ be a multiplicative character. If $\chi \neq \varepsilon$, then $\sum_t \chi(t) = 0$, where the sum is over all $t \in F_p$. If $\chi = \varepsilon$, the value of the sum is p.

PROOF

The last assertion is obvious, so we may assume that $\chi \neq \varepsilon$. In this case there is an $a \in F_p^*$ such that $\chi(a) \neq 1$. Let $T = \sum_t \chi(t)$. Then

$$\chi(a)T = \sum_t \chi(a)\chi(t) = \sum_t \chi(at) = T$$

The last equality follows since at runs over all elements of F_p as t does. Since $\chi(a)T = T$ and $\chi(a) \neq 1$ it follows that $T = 0$.

The multiplicative characters form a group by means of the following definitions. (We shall drop the use of the word *multiplicative* for the remainder of this chapter.)

1. If χ and λ are characters, then $\chi\lambda$ is the map that takes $a \in F_p^*$ to $\chi(a)\lambda(a)$.
2. If χ is a character, χ^{-1} is the map that takes $a \in F_p^*$ to $\chi(a)^{-1}$.

We leave it to the reader to verify that $\chi\lambda$ and χ^{-1} are characters and that these definitions make the set of characters into a group. The identity of this group is, of course, the trivial character ε.

Proposition 8.1.3

The group of characters is a cyclic group of order $p - 1$. If $a \in F_p^*$ and $a \neq 1$, then there is a character χ such that $\chi(a) \neq 1$.

PROOF

We know that F_p^* is cyclic (see Theorem 1 of Chapter 4). Let $g \in F_p^*$ be a generator. Then every $a \in F_p^*$ is equal to a power of g. If $a = g^l$ and χ is a character, then $\chi(a) = \chi(g)^l$. This shows that χ is completely determined by the value $\chi(g)$. Since $\chi(g)$ is a $(p - 1)$st root of unity, and since there are exactly $p - 1$ of these, it follows that the character group has order at most $p - 1$.

Now define a function λ by the equation $\lambda(g^k) = e^{2\pi i(k/(p-1))}$. It is easy to check that λ is well defined and is a character. We claim that $p - 1$ is the smallest integer n such that $\lambda^n = \varepsilon$. If $\lambda^n = \varepsilon$, then $\lambda^n(g) = \varepsilon(g) = 1$. However, $\lambda^n(g) = \lambda(g)^n = e^{2\pi i(n/(p-1))}$. It follows that $p - 1$ divides n. Since $\lambda^{p-1}(a) = \lambda(a)^{p-1} = \lambda(a^{p-1}) = \lambda(1) = 1$ we have $\lambda^{p-1} = \varepsilon$. We have established that the characters $\varepsilon, \lambda, \lambda^2, \ldots, \lambda^{p-2}$ are all distinct. Since by the first part of the proof there are at most $p - 1$ characters, we now have that there are exactly $p - 1$ characters and that the group is cyclic with λ as a generator.

If $a \in F_p^*$ and $a \neq 1$, then $a = g^l$ with $p - 1 \nmid l$. Let us compute $\lambda(a)$. $\lambda(a) = \lambda(g)^l = e^{2\pi i(l/(p-1))} \neq 1$. This concludes the proof.

Corollary

If $a \in F^*$ and $a \neq 1$, then $\sum_\chi \chi(a) = 0$, where the summation is over all characters.

PROOF

Let $S = \sum_\chi \chi(a)$. Since $a \neq 1$ there is, by the theorem, a character λ such that $\lambda(a) \neq 1$. Then

$$\lambda(a)S = \sum_\chi \lambda(a)\chi(a) = \sum_\chi \lambda\chi(a) = S$$

The final equality holds since $\lambda\chi$ runs over all characters as χ does. It follows that $(\lambda(a) - 1)S = 0$ and thus $S = 0$.

Characters are useful in the study of equations. To illustrate this, consider the equation $x^n = a$ for $a \in F_p^*$. By Proposition 4.2.1 we know that solutions exist iff $a^{(p-1)/d} = 1$, where $d = (n, p - 1)$, and that if a solution exists, then there are exactly d solutions. For simplicity, we shall assume that n divides $p - 1$. In this case $d = (n, p - 1) = n$.

We shall now derive a criterion for the solution of $x^n = a$ using characters.

Proposition 8.1.4

If $a \in F_p^*$, $n \mid p - 1$, and $x^n = a$ is not solvable, then there is a character χ such that

(a) $\chi^n = \varepsilon$.

(b) $\chi(a) \neq 1$.

PROOF

Let g and λ be as in Proposition 8.1.3 and set $\chi = \lambda^{(p-1)/n}$. Then $\chi(g) = \lambda^{(p-1)/n}(g) = \lambda(g)^{(p-1)/n} = e^{2\pi i/n}$. Now $a = g^l$ for some l, and since $x^n = a$ is not solvable, we must have $n \nmid l$. Then $\chi(a) = \chi(g)^l = e^{2\pi i(l/n)} \neq 1$. Finally, $\chi^n = \lambda^{p-1} = \varepsilon$.

For $a \in F_p$, let $N(x^n = a)$ denote the number of solutions of the equation $x^n = a$. If $n \mid p - 1$, we have

Proposition 8.1.5
$N(x^n = a) = \sum_{\chi^n = \varepsilon} \chi(a)$ where the sum is over all characters of order n.

PROOF
We claim first that there are exactly n characters of order n. Since the value of $\chi(g)$ for such a character must be an nth root of unity, there are at most n such characters. In Proposition 8.1.4, we found a character χ such that $\chi(g) = e^{2\pi i/n}$. It follows that $\varepsilon, \chi, \chi^2, \ldots, \chi^{n-1}$ are n distinct characters of order n.

To prove the formula, notice that $x^n = 0$ has one solution, namely, $x = 0$. Now $\sum_{\chi^n = \varepsilon} \chi(0) = 1$, since $\varepsilon(0) = 1$ and $\chi(0) = 1$ for $\chi \neq \varepsilon$.

Now suppose that $a \neq 0$ and that $x^n = a$ is solvable; i.e., there is an element b such that $b^n = a$. If $\chi^n = \varepsilon$, then $\chi(a) = \chi(b^n) = \chi(b)^n = \chi^n(b) = \varepsilon(b) = 1$. Thus $\sum_{\chi^n = \varepsilon} \chi(a) = n$, which is $N(x^n = a)$ in this case.

Finally, suppose that $a \neq 0$ and that $x^n = a$ is not solvable. We must show that $\sum_{\chi^n = \varepsilon} \chi(a) = 0$. Call the sum T. By Proposition 8.1.3, there is a character ρ such that $\rho(a) \neq 1$ and $\rho^n = \varepsilon$. A simple calculation shows that $\rho(a)T = T$ (one uses the obvious fact that the characters of order n form a group). Thus $(\rho(a) - 1)T = 0$ and $T = 0$, as required.

As a special case, suppose that p is odd and that $n = 2$. Then the theorem says that $N(x^2 = a) = 1 + (a/p)$, where (a/p) is the Legendre symbol. This equation is easy to check directly.

In Section 3 we shall return to equations over the field F_p.

2 *GAUSS SUMS*

In Chapter 6 we introduced quadratic Gauss sums. The following definition generalizes that notion.

Definition
Let χ be a character on F_p and $a \in F_p^*$. Set $g_a(\chi) = \sum_t \chi(t)\zeta^{at}$, where the sum is over all t in F_p, and $\zeta = e^{2\pi i/p}$. $g_a(\chi)$ is called a *Gauss sum* on F_p belonging to the character χ.

Proposition 8.2.1
If $a \neq 0$ and $\chi \neq \varepsilon$, we have $g_a(\chi) = \chi(a^{-1})g_1(\chi)$. If $a \neq 0$ and $\chi = \varepsilon$, we have $g_a(\varepsilon) = 0$. If $a = 0$ and $\chi \neq \varepsilon$, we have $g_0(\chi) = 0$. If $a = 0$ and $\chi = \varepsilon$, we have $g_0(\varepsilon) = p$.

PROOF

Suppose that $a \neq 0$ and that $\chi \neq \varepsilon$. Then

$$\chi(a)g_a(\chi) = \chi(a)\sum_t \chi(t)\zeta^{at} = \sum_t \chi(at)\zeta^{at} = g_1(\chi)$$

This proves the first assertion.

If $a \neq 0$, then

$$g_a(\varepsilon) = \sum_t \varepsilon(t)\zeta^{at} = \sum_t \zeta^{at} = 0$$

We have used Lemma 1 of Chapter 6.

To finish the proof notice that $g_0(\chi) = \sum_t \chi(t)\zeta^{0t} = \sum_t \chi(t)$. If $\chi = \varepsilon$, the result is p; if $\chi \neq \varepsilon$, the result is zero by Proposition 8.1.2.

From now on we shall denote $g_1(\chi)$ by $g(\chi)$. We wish to determine the absolute value of $g(\chi)$. This can be done fairly easily by imitating the proof of Proposition 6.3.2.

Proposition 8.2.2

If $\chi \neq \varepsilon$, then $|g(\chi)| = \sqrt{p}$.

PROOF

The idea is to evaluate the sum $\sum_a g_a(\chi)\overline{g_a(\chi)}$ in two ways.

If $a \neq 0$, then by Proposition 8.2.1 $\overline{g_a(\chi)} = \overline{\chi(a^{-1})g(\chi)} = \chi(a)\overline{g(\chi)}$ and $g_a(\chi) = \chi(a^{-1})g(\chi)$. Thus $g_a(\chi)\overline{g_a(\chi)} = \chi(a^{-1})\chi(a)g(\chi)\overline{g(\chi)} = |g(\chi)|^2$. Since $g_0(\chi) = 0$ our sum has the value $(p-1)|g(\chi)|^2$.

On the other hand,

$$g_a(\chi)\overline{g_a(\chi)} = \sum_x \sum_y \chi(x)\overline{\chi(y)}\zeta^{ax-ay}$$

Summing both sides over a and using the corollary to Lemma 1 of Chapter 6 yields

$$\sum_a g_a(\chi)\overline{g_a(x)} = \sum_x \sum_y \chi(x)\overline{\chi(y)}\delta(x, y)p = (p-1)p$$

Thus $(p-1)|g(\chi)|^2 = (p-1)p$ and the result follows.

The relation of the above result to Proposition 6.3.2 is made clearer by the following considerations.

What is the relation between $\overline{g(\chi)}$ and $g(\bar{\chi})$ [$\bar{\chi}$ is the character that takes a to $\overline{\chi(a)}$; i.e., it coincides with the character χ^{-1}]?

$$\overline{g(\chi)} = \sum_t \overline{\chi(t)}\zeta^{-t} = \chi(-1)\sum_t \overline{\chi(-t)}\zeta^{-t} = \chi(-1)g(\bar{\chi})$$

We have used the fact that $\overline{\chi(-1)} = \chi(-1)$, which is obvious since $\chi(-1) = \pm 1$. Thus the fact that $|g(\chi)|^2 = p$ can be written as $g(\chi)g(\bar{\chi}) = \chi(-1)p$. If χ is the Legendre symbol, this relation is precisely the result in Proposition 6.3.2.

3 *JACOBI SUMS*

Consider the equation $x^2 + y^2 = 1$ over the field F_p. Since F_p is finite, the equation has only finitely many solutions. Let $N(x^2 + y^2 = 1)$ be that number. We would like to determine this value explicitly.

Notice that

$$N(x^2 + y^2 = 1) = \sum_{a+b=1} N(x^2 = a)N(y^2 = b)$$

where the sum is over all pairs $a, b \in F_p$ such that $a + b = 1$. Since $N(x^2 = a) = 1 + (a/p)$, we obtain by substitution that

$$N(x^2 + y^2 = 1) = p + \sum_a (a/p) + \sum_b (b/p) + \sum_{a+b=1} (a/p)(b/p)$$

The first two sums are zero, so we are left with the task of evaluating the last sum. We shall see shortly that its value is $-(-1)^{(p-1)/2}$. Thus $N(x^2 + y^2 = 1)$ is $p - 1$ if $p \equiv 1\,(4)$ and $p + 1$ if $p \equiv 3\,(4)$. The reader is invited to check this result numerically for the first few primes.

Let us go a step further and try to evaluate $N(x^3 + y^3 = 1)$. As before we have

$$N(x^3 + y^3 = 1) = \sum_{a+b=1} N(x^3 = a)N(y^3 = b)$$

If $p \equiv 2\,(3)$, then $N(x^3 = a) = 1$ for all a since $(3, p - 1) = 1$. It follows that $N(x^3 + y^3 = 1) = p$ in this case. Assume now that $p \equiv 1\,(3)$. Let $\chi \neq \varepsilon$ be a character of order 3. Then χ^2 is a character of order 3 and $\chi^2 \neq \varepsilon$. Thus ε, χ, and χ^2 are all the characters of order 3, henceforth called cubic characters. By Proposition 8.1.5 we have $N(x^3 = a) = 1 + \chi(a) + \chi^2(a)$. Thus

$$N(x^3 + y^3 = 1) = \sum_{a+b=1} \sum_{i=0}^{2} \chi^i(a) \sum_{j=0}^{2} \chi^j(b)$$

$$= \sum_i \sum_j \left(\sum_{a+b=1} \chi^i(a)\chi^j(b) \right)$$

The inner sums are similar to the sum that occurred in the analysis of $N(x^2 + y^2 = 1)$.

Definition
Let χ and λ be characters of F_p and set $J(\chi, \lambda) = \sum_{a+b=1} \chi(a)\lambda(b)$. $J(\chi, \lambda)$ is called a *Jacobi sum.*

To complete the analysis of $N(x^2 + y^2 = 1)$ and $N(x^3 + y^3 = 1)$ we need to obtain information on the value of Jacobi sums. The following theorem not only supplies this information, but shows as well a surprising connection between Jacobi sums and Gauss sums.

Theorem 1
Let χ and λ be nontrivial characters. Then

(a) $J(\varepsilon, \varepsilon) = p$.
(b) $J(\varepsilon, \chi) = 0$.
(c) $J(\chi, \chi^{-1}) = -\chi(-1)$.
(d) *If $\chi\lambda \neq \varepsilon$, then*

$$J(\chi, \lambda) = \frac{g(\chi)g(\lambda)}{g(\chi\lambda)}$$

PROOF
Part (a) is immediate, and part (b) is an immediate consequence of Proposition 8.1.2.

To prove part (c), notice that

$$J(\chi, \chi^{-1}) = \sum_{a+b=1} \chi(a)\chi^{-1}(b) = \sum_{\substack{a+b=1 \\ b\neq 0}} \chi\left(\frac{a}{b}\right) = \sum_{a\neq 1} \chi\left(\frac{a}{1-a}\right)$$

Set $a/(1-a) = c$. If $c \neq -1$, then $a = c/(1+c)$. It follows that as a varies over F_p, less the element 1, that c varies over F_p, less the element -1. Thus

$$J(\chi, \chi^{-1}) = \sum_{c\neq -1} \chi(c) = -\chi(-1)$$

To prove part (d), notice that

$$\begin{aligned} g(\chi)g(\lambda) &= \left(\sum_x \chi(x)\zeta^x\right)\left(\sum_y \lambda(y)\zeta^y\right) \\ &= \sum_{x,y} \chi(x)\lambda(y)\zeta^{x+y} \\ &= \sum_t \left(\sum_{x+y=t} \chi(x)\lambda(y)\right)\zeta^t \end{aligned} \tag{1}$$

If $t = 0$, then $\sum_{x+y=0} \chi(x)\lambda(y) = \sum_x \chi(x)\lambda(-x) = \lambda(-1)\sum_x \chi\lambda(x) = 0$, since $\chi\lambda \neq \varepsilon$ by assumption.

If $t \neq 0$, define x' and y' by $x = tx'$ and $y = ty'$. If $x + y = t$, then $x' + y' = 1$. It follows that

$$\sum_{x+y=t} \chi(x)\lambda(y) = \sum_{x'+y'=1} \chi(tx')\lambda(ty') = \chi\lambda(t)J(\chi, \lambda)$$

Substituting into Equation (1) yields

$$g(\chi)g(\lambda) = \sum_t \chi\lambda(t)J(\chi, \lambda)\zeta^t = J(\chi, \lambda)g(\chi\lambda)$$

Corollary

If χ, λ, and $\chi\lambda$ are not equal to ε, then $|J(\chi, \lambda)| = \sqrt{p}$.

PROOF

Take the absolute value of both sides of the equation in part (d) and use Proposition 8.2.2.

We now return to the analysis of $N(x^2 + y^2 = 1)$ and $N(x^3 + y^3 = 1)$. In the former case, it was necessary to evaluate the sum $\sum_{a+b=1}(a/p) \times (b/p)$. Case (c) of Theorem 1 is applicable and gives the result $-(-1/p) = -(-1)^{(p-1)/2}$, as was stated earlier.

In the case of $N(x^3 + y^3 = 1)$ we had to evaluate the sums $\sum_{a+b=1}\chi^i(a)\chi^j(b)$, where χ is a cubic character. Applying the theorem leads to the result

$$N(x^3 + y^3 = 1) = p - \chi(-1) - \chi^2(-1) + J(\chi, \chi) + J(\chi^2, \chi^2)$$

Since $-1 = (-1)^3$ we have $\chi(-1) = \chi^2(-1) = 1$. Also notice that $\chi^2 = \chi^{-1} = \bar{\chi}$. Thus

$$N(x^3 + y^3 = 1) = p - 2 + 2\,\mathrm{Re}\, J(\chi, \chi)$$

This result is not as nice as the result for $N(x^2 + y^2 = 1)$, since we do not know $J(\chi, \chi)$ explicitly. Nevertheless, by the corollary to Theorem 1 we know that $|J(\chi, \chi)| = \sqrt{p}$ so we have the estimate

$$|N(x^3 + y^3 = 1) - p + 2| \leq 2\sqrt{p}$$

If we write N_p for the number of solutions to $x^3 + y^3 = 1$ in the field F_p, then the estimate says that N_p is approximately equal to $p - 2$ with an "error term" $2\sqrt{p}$. This shows that for large primes p there are always many solutions.

If $p \equiv 1$ (3), there are always at least six solutions since $x^3 = 1$ and $y^3 = 1$ have three solutions each and we can write $1 + 0 = 1$ and $0 + 1 = 1$. For $p = 7$ and 13 these are the only solutions. For $p = 19$ other solutions exist; e.g., $3^3 + 10^3 \equiv 1$ (19). These "nontrivial" solutions exist for all primes $p \geq 19$ since it follows from the estimate that $N_p \geq p - 2 - 2\sqrt{p} > 6$ for $p \geq 19$.

Using Jacobi sums we can easily extend our analysis to equations of the form $ax^n + by^n = 1$, but we shall not go more deeply into this matter now.

The corollary to Theorem 1 has two immediate consequences of considerable interest.

Proposition 8.3.1

If $p \equiv 1\ (4)$, *then there exist integers a and b such that* $a^2 + b^2 = p$.
If $p \equiv 1\ (3)$, *then there exist integers a and b such that* $a^2 - ab + b^2 = p$.

PROOF

If $p \equiv 1\ (4)$, there is a character χ of order 4 (if λ has order $p - 1$, let $\chi = \lambda^{(p-1)/4}$). The values of χ are in the set $\{1, -1, i, -i\}$, where $i = \sqrt{-1}$. Thus $J(\chi, \chi) = \sum_{s+t=1} \chi(s)\chi(t) \in Z[i]$ (see Chapter 1, Section 4). It follows that $J(\chi, \chi) = a + bi$, where $a, b \in Z$; thus $p = |J(\chi, \chi)|^2 = a^2 + b^2$.

If $p \equiv 1\ (3)$, there is a character χ of order 3. The values of χ are in the set $\{1, \omega, \omega^2\}$, where $\omega = e^{2\pi i/3} = (-1 + \sqrt{-3})/2$. Thus $J(\chi, \chi) \in Z[\omega]$. As above, we have $J(\chi, \chi) = a + b\omega$, where $a, b \in Z$ and $p = |J(\chi, \chi)|^2 = |a + b\omega|^2 = a^2 - ab + b^2$.

The fact that primes $p \equiv 1\ (4)$ can be written as the sum of two squares was discovered by Fermat. It is not hard to prove that if $a, b > 0$, a is odd and b is even, then the representation $p = a^2 + b^2$ is unique.

If $p \equiv 1\ (3)$, the representation $p = a^2 - ab + b^2$ is not unique even if we assume that $a, b > 0$. This can be seen from the equations

$$a^2 - ab + b^2 = (b - a)^2 - (b - a)b + b^2 = a^2 - a(a - b) + (a - b)^2$$

However, we can reformulate things so that the result is unique. If $p = a^2 - ab + b^2$, then $4p = (2a - b)^2 + 3b^2 = (2b - a)^2 + 3a^2 = (a + b)^2 + 3(a - b)^2$. We claim that 3 divides either a, b, or $a - b$. Suppose that $3 \nmid a$ and that $3 \nmid b$. If $a \equiv 1\ (3)$ and $b \equiv 2\ (3)$, or $a \equiv 2\ (3)$ and $b \equiv 1\ (3)$, then $a^2 - ab + b^2 \equiv 0\ (3)$, which implies that $3 \mid p$, a contradiction. Thus $3 \mid a - b$, and we have

Proposition 8.3.2

If $p \equiv 1\ (3)$, *then there are integers A and B such that* $4p = A^2 + 27B^2$. *In this representation of* $4p$, *A and B are uniquely determined up to sign.*

PROOF

The proof of the uniqueness is left to the Exercises.

Theorem 1 together with a simple argument leads to a further interesting relation between Gauss sums and Jacobi sums.

Proposition 8.3.3
Suppose that $p \equiv 1\ (n)$ and that χ is a character of order n. Then

$$g(\chi)^n = \chi(-1)pJ(\chi, \chi)J(\chi, \chi^2)\cdots J(\chi, \chi^{n-2})$$

PROOF
Using part (d) of Theorem 1 we have $g(\chi)^2 = J(\chi, \chi)g(\chi^2)$. Multiply both sides by $g(\chi)$ and we get $g(\chi)^3 = J(\chi, \chi)J(\chi, \chi^2)g(\chi^3)$. Continuing in this way shows that

$$g(\chi)^{n-1} = J(\chi, \chi)J(\chi, \chi^2)\cdots J(\chi, \chi^{n-2})g(\chi^{n-1}) \tag{2}$$

Now $\chi^{n-1} = \chi^{-1} = \bar{\chi}$. Thus, as we have seen, $g(\chi)g(\chi^{n-1}) = g(\chi)g(\bar{\chi}) = \chi(-1)p$. The result follows upon multiplying both sides of Equation (2) by $g(\chi)$.

Corollary
If χ is a cubic character, then

$$g(\chi)^3 = pJ(\chi, \chi)$$

PROOF
This is simply a special case of the proposition and the fact that $\chi(-1) = \chi((-1)^3) = 1$.

Using this corollary, we are in a position to analyze more fully the complex number $J(\chi, \chi)$ that occurred in the discussion of $N(x^3 + y^3 = 1)$. We have seen that $J(\chi, \chi) = a + b\omega$, where $a, b \in Z$ and $\omega = e^{2\pi i/3} = (-1 + \sqrt{-3})/2$.

Proposition 8.3.4
Suppose that $p \equiv 1\ (3)$ and that χ is a cubic character. Set $J(\chi, \chi) = a + b\omega$ as above. Then

(a) $b \equiv 0\ (3)$.
(b) $a \equiv -1\ (3)$.

PROOF
We shall work with congruences in the ring of algebraic integers as in Chapter 6:

$$g(\chi)^3 = \left(\sum_t \chi(t)\zeta^t\right)^3 \equiv \sum_t \chi(t)^3\zeta^{3t} \tag{3}$$

Since $\chi(0) = 0$ and $\chi(t)^3 = 1$ for $t \neq 0$ we have $\sum_t \chi(t)^3 \zeta^{3t} = \sum_{t \neq 0} \zeta^{3t} = -1$. Thus

$$g(\chi)^3 = pJ(\chi, \chi) \equiv a + b\omega \equiv -1 \ (3)$$

Working with $\bar{\chi}$ instead of χ and remembering that $\overline{g(\chi)} = g(\bar{\chi})$ we find that

$$g(\bar{\chi})^3 = pJ(\bar{\chi}, \bar{\chi}) \equiv a + b\bar{\omega} \equiv -1 \ (3)$$

Subtracting yields $b(\omega - \bar{\omega}) \equiv 0$ (3), or $b\sqrt{-3} \equiv 0$ (3). Thus $-3b^2 \equiv 0$ (9) and it follows that $3 \mid b$. Since $3 \mid b$ and $a + b\omega \equiv -1$ (3), we must have $a \equiv -1$ (3), which completes the proof.

Corollary

Let $A = 2a - b$ and $B = b/3$. Then $A \equiv 1$ (3) and

$$4p = A^2 + 27B^2$$

PROOF

Since $J(\chi, \chi) = a + b\omega$ and $|J(\chi, \chi)|^2 = p$ we have $p = a^2 - ab + b^2$. Thus $4p = (2a - b)^2 + 3b^2$ and $4p = A^2 + 27B^2$.

By Proposition 8.3.4, $3 \mid b$ and $a \equiv -1$ (3). Therefore, $A = 2a - b \equiv 1$ (3).

We are now ready to prove the following beautiful theorem due to Gauss.

Theorem 2

Suppose that $p \equiv 1$ (3). Then there are integers A and B such that $4p = A^2 + 27B^2$. If we require that $A \equiv 1$ (3), A is uniquely determined, and

$$N(x^3 + y^3 = 1) = p - 2 + A$$

PROOF

We have already shown that $N(x^3 + y^3 = 1) = p - 2 + 2 \operatorname{Re} J(\chi, \chi)$. Since $J(\chi, \chi) = a + b\omega$ as above, we have $\operatorname{Re} J(\chi, \chi) = (2a - b)/2$. Thus $2 \operatorname{Re} J(\chi, \chi) = 2a - b = A \equiv 1$ (3).

Let us illustrate this result with two examples, $p = 61$ and $p = 67$.

$4 \cdot 61 = 1^2 + 27 \cdot 3^2$. Thus the number of solutions to $x^3 + y^3 = 1$ in F_{61} is $61 - 2 + 1 = 60$.

Now, $4 \cdot 67 = 5^2 + 27 \cdot 3^2$. We must be careful here; since $5 \not\equiv 1$ (3) we must choose $A = -5$. The answer is thus $67 - 2 - 5 = 60$, which by coincidence (?) is the same as for $p = 61$.

4 THE EQUATION $x^n + y^n = 1$ IN F_p

We shall assume that $p \equiv 1\ (n)$ and investigate the number of solutions to the equation $x^n + y^n = 1$ over the field F_p. The methods of Section 3 are directly applicable.

We have

$$N(x^n + y^n = 1) = \sum_{a+b=1} N(x^n = a)N(x^n = b)$$

Let χ be a character of order n. By Proposition 8.1.5

$$N(x^n = a) = \sum_{i=0}^{n-1} \chi^i(a)$$

Combining these results yields

$$N(x^n + y^n = 1) = \sum_{j=0}^{n-1} \sum_{i=0}^{n-1} J(\chi^j, \chi^i)$$

Theorem 1 can be used to estimate this sum. When $i = j = 0$ we have $J(\chi^0, \chi^0) = J(\varepsilon, \varepsilon) = p$. When $j + i = n$, $\chi^j = (\chi^i)^{-1}$ so that $J(\chi^j, \chi^i) = -\chi^j(-1)$. The sum of these terms is $-\sum_{j=1}^{n-1} \chi^j(-1)$. $\sum_{j=0}^{n-1} \chi^j(-1)$ is n when -1 is an nth power and zero otherwise. Thus the contribution of these terms is $1 - \delta_n(-1)n$, where $\delta_n(-1)$ has the obvious meaning. Finally, if $i = 0$ and $j \neq 0$ or $i \neq 0$ and $j = 0$, then $J(\chi^i, \chi^j) = 0$. Thus

$$N(x^n + y^n = 1) = p + 1 - \delta_n(-1)n + \sum_{i,j} J(\chi^i, \chi^j)$$

The sum is over indices i and j between 1 and $n - 1$ subject to the condition that $i + j \neq n$. There are $(n-1)^2 - (n-1) = (n-1)(n-2)$ such terms and they all have absolute value $\sqrt{p}$. Thus

Proposition 8.4.1

$$|N(x^n + y^n = 1) + \delta_n(-1)n - (p+1)| \leq (n-1)(n-2)\sqrt{p}$$

The term $\delta_n(-1)n$ will be interpreted later as the number of points "at infinity" on the curve $x^n + y^n = 1$.

For large p the above estimate shows the existence of many nontrivial solutions.

5 MORE ON JACOBI SUMS

Theorem 1 can be generalized in a very fruitful manner. First we need a definition.

Definition

Let $\chi_1, \chi_2, \ldots, \chi_l$ be characters on F_p. A Jacobi sum is defined by the formula

$$J(\chi_1, \chi_2, \ldots, \chi_l) = \sum_{t_1 + \cdots + t_l = 1} \chi_1(t_1)\chi_2(t_2)\cdots\chi_l(t_l)$$

Notice that when $l = 2$ this reduces to our former definition of Jacobi sum.

It is useful to define another sum, which will be left unnamed:

$$J_0(\chi_1, \ldots, \chi_l) = \sum_{t_1 + \cdots + t_l = 0} \chi_1(t_1)\chi_2(t_2)\cdots\chi_l(t_l)$$

Proposition 8.5.1

(a) $J_0(\varepsilon, \varepsilon, \ldots, \varepsilon) = J(\varepsilon, \varepsilon, \ldots, \varepsilon) = p^{l-1}$.

(b) *If some but not all of the* χ_i *are trivial, then* $J_0(\chi_1, \chi_2, \ldots, \chi_l) = J(\chi_1, \chi_2, \ldots, \chi_l) = 0$.

(c) *Assume that* $\chi_l \neq \varepsilon$. *Then*

$$J_0(\chi_1, \chi_2, \ldots, \chi_l) = \begin{cases} 0 & \text{if } \chi_1\chi_2\cdots\chi_l \neq \varepsilon \\ \chi_l(-1)(p-1)J(\chi_1, \chi_2, \ldots, \chi_{l-1}) & \textit{otherwise} \end{cases}$$

PROOF

If $t_1, t_2, \ldots, t_{l-1}$ are chosen (arbitrarily) in F_p, then t_l is uniquely determined by the condition $t_1 + t_2 + \cdots + t_{l-1} + t_l = 0$. Thus $J_0(\varepsilon, \varepsilon, \ldots, \varepsilon) = p^{l-1}$. Similarly for $J(\varepsilon, \varepsilon, \ldots, \varepsilon)$.

To prove part (b), assume that $\chi_1, \chi_2, \ldots, \chi_s$ are nontrivial and that $\chi_{s+1} = \chi_{s+2} = \cdots = \chi_l = \varepsilon$. Then

$$\sum_{t_1 + \cdots + t_l = 0} \chi_1(t_1)\chi_2(t_2)\cdots\chi_l(t_l)$$

$$= \sum_{t_1, t_2, \ldots, t_{l-1}} \chi_1(t_1)\chi_2(t_2)\cdots\chi_s(t_s)$$

$$= p^{l-s-1}\left(\sum_{t_1}\chi_1(t_1)\right)\left(\sum_{t_2}\chi_2(t_2)\right)\cdots\left(\sum_{t_s}\chi_s(t_s)\right) = 0$$

We have used Proposition 8.1.2. Thus $J_0(\chi_1, \chi_2, \ldots, \chi_l) = 0$. Similarly for $J(\chi_1, \ldots, \chi_l)$.

To prove part (c), notice that

$$J_0(\chi_1, \chi_2, \ldots, \chi_l) = \sum_s \left(\sum_{t_1 + \cdots + t_{l-1} = -s} \chi_1(t_1) \cdots \chi_{l-1}(t_{l-1}) \right) \chi_l(s)$$

Since $\chi_l \neq \varepsilon$, $\chi_l(0) = 0$, so we may assume that $s \neq 0$ in the above sum. If $s \neq 0$, define t_i' by $t_i = -st_i'$. Then

$$\sum_{t_1 + \cdots + t_{l-1} = -s} \chi_1(t_1) \cdots \chi_{l-1}(t_{l-1})$$

$$= \chi_1\chi_2 \cdots \chi_{l-1}(-s) \sum_{t_1' + \cdots + t_{l-1}' = 1} \chi_1(t_1') \cdots \chi_{l-1}(t_{l-1}')$$

$$= \chi_1\chi_2 \cdots \chi_{l-1}(-s)J(\chi_1, \ldots, \chi_{l-1})$$

Combining these results yields

$$J_0(\chi_1, \chi_2, \ldots, \chi_l) = \chi_1\chi_2 \cdots \chi_{l-1}(-1)J(\chi_1, \ldots, \chi_{l-1}) \sum_{s \neq 0} \chi_1\chi_2 \cdots \chi_l(s)$$

The main result follows since the sum is zero if $\chi_1\chi_2 \cdots \chi_l \neq \varepsilon$ and $p - 1$ if $\chi_1\chi_2 \cdots \chi_l = \varepsilon$.

Parts (a) and (b) of Proposition 8.5.1 generalize parts (a) and (b) of Theorem 1. Part (d) of Theorem 1 can be generalized as follows.

Theorem 3

Assume that $\chi_1, \chi_2, \ldots, \chi_r$ are nontrivial and also that $\chi_1\chi_2 \cdots \chi_r$ is nontrivial. Then

$$g(\chi_1)g(\chi_2) \cdots g(\chi_r) = J(\chi_1, \chi_2, \ldots, \chi_r)g(\chi_1\chi_2 \cdots \chi_r)$$

PROOF

Let $\psi : F_p \to \mathbb{C}$ be defined by $\psi(t) = \zeta^t$. Then $\psi(t_1 + t_2) = \psi(t_1)\psi(t_2)$, and $g(\chi) = \sum \chi(t)\psi(t)$. The introduction of ψ is for notational convenience.

$$g(\chi_1)g(\chi_2) \cdots g(\chi_r)$$

$$= \left(\sum_{t_1} \chi_1(t_1)\psi(t_1) \right) \cdots \left(\sum_{t_r} \chi_r(t_r)\psi(t_r) \right)$$

$$= \sum_s \left(\sum_{t_1 + t_2 + \cdots + t_r = s} \chi_1(t_1)\chi_2(t_2) \cdots \chi_r(t_r) \right) \psi(s)$$

If $s = 0$, then by part (c) of Proposition 8.5.1 and the assumption that $\chi_1 \cdots \chi_r \neq \varepsilon$

$$\sum_{t_1 + \cdots + t_r = 0} \chi_1(t_1) \cdots \chi_r(t_r) = 0$$

If $s \neq 0$, the substitution $t_i' = st_i$ shows that

$$\sum_{t_1+\cdots+t_r=s} \chi_1(t_1)\cdots\chi_r(t_r) = \chi_1\chi_2\cdots\chi_r(s)J(\chi_1, \chi_2, \ldots, \chi_r)$$

Putting these remarks together, we have

$$\begin{aligned} g(\chi_1)\cdots g(\chi_r) &= J(\chi_1, \chi_2, \ldots, \chi_r) \sum_{s\neq 0} \chi_1\chi_2\cdots\chi_r(s)\psi(s) \\ &= J(\chi_1, \chi_2, \ldots, \chi_r)g(\chi_1\chi_2\cdots\chi_r) \end{aligned}$$

Corollary 1

Suppose that $\chi_1, \chi_2, \ldots, \chi_r$ are nontrivial and that $\chi_1\chi_2\cdots\chi_r$ is trivial. Then

$$g(\chi_1)g(\chi_2)\cdots g(\chi_r) = \chi_r(-1)pJ(\chi_1, \chi_2, \ldots, \chi_{r-1})$$

PROOF

$g(\chi_1)g(\chi_2)\cdots g(\chi_{r-1}) = J(\chi_1, \ldots, \chi_{r-1})g(\chi_1\chi_2\cdots\chi_{r-1})$ by Theorem 3. Multiply both sides by $g(\chi_r)$. Since $\chi_1\chi_2\cdots\chi_{r-1} = \chi_r^{-1}$ we have

$$g(\chi_1\cdots\chi_{r-1})g(\chi_r) = g(\chi_r^{-1})g(\chi_r) = \chi_r(-1)p$$

Corollary 2

Let the hypotheses be as in Corollary 1. Then

$$J(\chi_1, \ldots, \chi_r) = -\chi_r(-1)J(\chi_1, \chi_2, \ldots, \chi_{r-1})$$

[If $r = 2$, we set $J(\chi_1) = 1$.]

PROOF

If $r = 2$, this is the assertion of part (c) of Theorem 1.

Suppose that $r > 2$. In the proof of Theorem 3 use the hypothesis that $\chi_1\chi_2\cdots\chi_r = \varepsilon$. This yields

$$g(\chi_1)g(\chi_2)\cdots g(\chi_r) = J_0(\chi_1, \chi_2, \ldots, \chi_r) + J(\chi_1, \ldots, \chi_r) \sum_{s\neq 0} \psi(s)$$

Since $\sum_s \psi(s) = 0$, the sum in the formula is equal to -1. By part (c) of Proposition 8.5.1, we have $J_0(\chi_1, \ldots, \chi_r) = \chi_r(-1)(p-1)J(\chi_1, \ldots, \chi_{r-1})$. By Corollary 1, $g(\chi_1)\cdots g(\chi_r) = \chi_r(-1)pJ(\chi_1, \chi_2, \ldots, \chi_{r-1})$. Putting these results together proves the corollary.

Theorem 4

Assume that $\chi_1, \chi_2, \ldots, \chi_r$ are nontrivial.

(a) *If $\chi_1\chi_2\cdots\chi_r \neq \varepsilon$, then*

$$|J(\chi_1, \chi_2, \ldots, \chi_r)| = p^{(r-1)/2}$$

(b) *If* $\chi_1\chi_2\cdots\chi_r = \varepsilon$, *then*

$$|J_0(\chi_1, \chi_2, \ldots, \chi_r)| = (p-1)p^{(r/2)-1}$$

and

$$|J(\chi_1, \chi_2, \ldots, \chi_r)| = p^{(r/2)-1}$$

PROOF

If χ is nontrivial, $|g(\chi)| = \sqrt{p}$. Part (a) follows directly from Theorem 3.

Part (b) follows similarly from part (c) of Proposition 8.5.1 and from Corollary 2 to Theorem 3.

6 *APPLICATIONS*

Earlier in this chapter we investigated the number of solutions of the equation $x^2 + y^2 = 1$ in the field F_p. It is natural to ask the same question about the equation $x_1^2 + x_2^2 + \cdots + x_r^2 = 1$. The answer can easily be found using the results of Section 5.

Let χ be a character of order 2 ($\chi(a) = (a/p)$ in our earlier notation). Then $N(x^2 = a) = 1 + \chi(a)$. Thus

$$N(x_1^2 + \cdots + x_r^2 = 1) = \sum N(x_1^2 = a_1)N(x_2^2 = a_2)\cdots N(x_r^2 = a_r)$$

where the sum is over all r-tuples $(a_1, \ldots, a_r)$ such that $a_1 + a_2 + \cdots + a_r = 1$. Multiplying out, and using Proposition 8.5.1, yields

$$N(x_1^2 + \cdots + x_r^2 = 1) = p^{r-1} + J(\chi, \chi, \ldots, \chi)$$

If r is odd, $\chi^r = \chi$, and if r is even, $\chi^r = \varepsilon$.

Suppose that r is odd. Then Theorem 3 applies and we have $J(\chi, \ldots, \chi) = g(\chi)^{r-1}$. Since $g(\chi)^2 = \chi(-1)p$ it follows that $J(\chi, \ldots, \chi) = \chi(-1)^{(r-1)/2}p^{(r-1)/2}$.

If r is even, we use Corollary 2 to Theorem 3 and find that $J(\chi, \chi, \ldots, \chi) = -\chi(-1)^{r/2}p^{(r-2)/2}$. Finally, remember that $\chi(-1) = (-1)^{(p-1)/2}$. Thus

Proposition 8.6.1

If r is odd, then

$$N(x_1^2 + x_2^2 + \cdots + x_r^2 = 1) = p^{r-1} + (-1)^{((r-1)/2)((p-1)/2)}p^{(r-1)/2}$$

If r is even, then

$$N(x_1^2 + x_2^2 + \cdots + x_r^2 = 1) = p^{r-1} - (-1)^{(r/2)((p-1)/2)}p^{(r/2)-1}$$

The most general equation that can be treated by these methods has the form $a_1x_1^{l_1} + a_2x_2^{l_2} + \cdots + a_rx_r^{l_r} = b$, where $a_1, \ldots, a_r, b \in F_p$, and $l_1, l_2, \ldots, l_r$ are positive integers. We shall return to this subject in Section 7. For now, we shall use Jacobi sums to give yet another proof of the law of quadratic reciprocity.

Let q be an odd prime not equal to p, and χ the character of order 2 on F_p. Then by Corollary 1 to Theorem 3

$$g(\chi)^{q+1} = (-1)^{(p-1)/2} pJ(\chi, \chi, \ldots, \chi)$$

where there are q components in the Jacobi sum.

Since $q + 1$ is even $g(\chi)^{q+1} = (g(\chi)^2)^{(q+1)/2} = (-1)^{((p-1)/2)((q+1)/2)} \cdot p^{(q+1)/2}$. Substituting into the formula we find that

$$(-1)^{((p-1)/2)((q-1)/2)} p^{(q-1)/2} = J(\chi, \chi, \ldots, \chi)$$

Now, $J(\chi, \chi, \ldots, \chi) = \sum \chi(t_1)\chi(t_2) \cdots \chi(t_q)$, where the sum is over all $(t_1, t_2, \ldots, t_q)$ with $t_1 + t_2 + \cdots + t_q = 1$. If $t = t_1 = t_2 = \cdots = t_q$, then $t = 1/q$, and the corresponding term of the sum has value $\chi(1/q)^q = \chi(q)^{-q} = \chi(q)$. If not all the t_i are equal, then there are q different q-tuples obtained from $(t_1, t_2, \ldots, t_q)$ by cyclic permutation. The corresponding terms of the sum all have the same value. Thus

$$(-1)^{((p-1)/2)((q-1)/2)} p^{(q-1)/2} \equiv \chi(q) \ (q)$$

Since $\chi(q) = (q/p)$ and $p^{(q-1)/2} \equiv (p/q) \ (q)$ we have

$$(-1)^{((p-1)/2)((q-1)/2)} (p/q) \equiv (q/p) \ (q)$$

and thus

$$(-1)^{((p-1)/2)((q-1)/2)} (p/q) = (q/p)$$

7 A GENERAL THEOREM

All the equations we have considered up to now are special cases of

$$a_1 x_1^{l_1} + a_2 x_2^{l_2} + \cdots + a_r x_r^{l_r} = b \tag{3}$$

where $a_1, a_2, \ldots, a_r, b \in F_p^*$. Let N be the number of solutions. Our object is to give a formula for N and an estimate for N. The methods to be used are identical with those already developed in the previous sections.

To begin with, we have

$$N = \sum N(x_1^{l_1} = u_1) N(x_2^{l_2} = u_2) \cdots N(x_r^{l_r} = u_r) \tag{4}$$

where the sum is over all r-tuples $(u_1, u_2, \ldots, u_r)$ such that $\sum_{i=1}^{r} a_i u_i = b$.

We shall assume that $l_1, l_2, \ldots, l_r$ are divisors of $p - 1$, although this is not necessary (see the Exercises). Let χ_i vary over the characters of order l_i. Then

$$N(x_i^{l_i} = u_i) = \sum_{\chi_i} \chi_i(u_i)$$

Substituting into Equation (4) we get

$$N = \sum_{\chi_1, \chi_2, \ldots, \chi_r} \sum_{\Sigma a_i u_i = b} \chi_1(u_1)\chi_2(u_2) \cdots \chi_r(u_r) \tag{5}$$

The inner sum is closely related to the Jacobi sums that we have considered.

It is necessary to treat the cases $b = 0$ and $b \neq 0$ separately.

If $b = 0$, let $t_i = a_i u_i$. Then the inner sum becomes

$$\chi_1(a_1^{-1})\chi_2(a_2^{-1}) \cdots \chi_r(a_r^{-1}) J_0(\chi_1, \chi_2, \ldots, \chi_r)$$

If $b \neq 0$, let $t_i = b^{-1} a_i u_i$. The inner sum becomes

$$\chi_1 \chi_2 \cdots \chi_r(b) \chi_1(a_1^{-1}) \cdots \chi_r(a_r^{-1}) J(\chi_1, \chi_2, \ldots, \chi_r)$$

In both cases, if $\chi_1 = \chi_2 = \cdots = \chi_r = \varepsilon$, the term has the value p^{r-1} since $J_0(\varepsilon, \ldots, \varepsilon) = J(\varepsilon, \varepsilon, \ldots, \varepsilon) = p^{r-1}$. If some but not all the χ_i are equal to ε, then the term has the value zero. In the first case the value is zero unless $\chi_1 \chi_2 \cdots \chi_r = \varepsilon$. All this is a consequence of Proposition 8.5.1.

Putting this together with Theorem 4 we obtain

Theorem 5

If $b = 0$, then

$$N = p^{r-1} + \sum \chi_1(a_1^{-1})\chi_2(a_2^{-1}) \cdots \chi_r(a_r^{-1}) J_0(\chi_1, \chi_2, \ldots, \chi_r)$$

The sum is over all r-tuples of characters $\chi_1, \chi_2, \ldots, \chi_r$, where $\chi_i^{l_i} = \varepsilon$, $\chi_i \neq \varepsilon$ for $i = 1, \ldots, r$, and $\chi_1 \chi_2 \cdots \chi_r = \varepsilon$. If M is the number of such r-tuples, then

$$|N - p^{r-1}| \leq M(p-1)p^{(r/2)-1}$$

If $b \neq 0$, then

$$N = p^{r-1} + \sum \chi_1 \chi_2 \cdots \chi_r(b) \chi_1(a_1^{-1}) \cdots \chi_r(a_r^{-1}) J(\chi_1, \chi_2, \ldots, \chi_r)$$

The summation is over all r-tuples of characters $\chi_1, \ldots, \chi_r$, where $\chi_i^{l_i} = \varepsilon$ and $\chi_i \neq \varepsilon$ for $i = 1, \ldots, r$. If M_0 is the number of such r-tuples with $\chi_1 \chi_2 \cdots \chi_r = \varepsilon$, and M_1 is the number of such r-tuples with $\chi_1 \chi_2 \cdots \chi_r \neq \varepsilon$, then

$$|N - p^{r-1}| \leq M_0 p^{(r/2)-1} + M_1 p^{(r-1)/2}$$

An immediate consequence of Theorem 5 is worth noting. Let $a_1, a_2, \ldots, a_r$ and $b \in \mathbb{Z}$ and consider the congruence

$$a_1 x_1^{l_1} + a_2 x_2^{l_2} + \cdots + a_r x_r^{l_r} \equiv b\ (p)$$

Then if p is sufficiently large, the congruence has many solutions. In fact, the number of solutions tends to infinity as p is taken larger and larger.

Notes

The inspiration for this chapter and much that follows is the famous paper of Weil [80]. The basic relationship between Gauss and Jacobi sums was known to Gauss, Jacobi, and Eisenstein. Aside from its usefulness in obtaining the Weil–Riemann hypothesis for certain hypersurfaces over finite fields (see Chapter 11) the generalized Jacobi sum is of importance in the theory of cyclotomy and difference sets. For an introduction to this material, see Storer [74]. See also the difficult but important continuation of [80] by Weil [81].

Material on Gauss and Jacobi sums is scattered throughout the treatise of Hasse [41]. He gives a systematic presentation in his last chapter where in addition to developing many interesting results he shows how both types of sum arise naturally in the theory of cyclotomic number fields. Much of the theory in that chapter is distilled from the paper of Davenport and Hasse [23]. The latter paper is well worth close study, but it is unfortunately of an advanced nature and is probably inaccessible to a beginner. Somewhat less difficult are the more recent papers of K. Yamamoto [82] and A. Yokoyama [83]. One should also consult the classical treatise of P. Bachman [5].

Theorem 2 is proved by Gauss in Article 358 of *Disquisitiones Arithmeticae*. He does not really state the theorem explicitly. It comes out as a by-product of another investigation. What he does, in fact, is to use the theorem to help find the algebraic equation satisfied by certain Gauss sums. We have done the reverse, using the theory of Gauss sums to derive the theorem. Gauss derived other results of this type in his first memoir on biquadratic reciprocity [34]. For further historical remarks about this subject, see the introduction to the paper of Weil [80].

The estimates given in Theorem 5 are derived in the first chapter of Borevich and Shafarevich [9]. They use a somewhat different method, which we have outlined in the Exercises. In the special case of quadratic forms, i.e., when all the $l = 2$, the result goes back at least to Dickson [25].

The technique of counting solutions by means of characters lends itself naturally to the problem of finding sequences of integers of prescribed length having prescribed kth power character modulo p. This problem is dealt with to some extent in Hasse [41]. In an interesting, and elementary, paper, Davenport [21] shows that the number of sequences of four successive quadratic residues between 1 and p satisfies the inequality $|R - p/8| < Kp^{3/4}$, where K is a constant independent of p. Better estimates can be obtained using the results of Weil. For another paper along the same lines, see Graham [36].

One final remark on Theorem 5. It is due originally to Weil and independently (and almost simultaneously) to L. K. Hua and H. S. Vandiver (*Proc. Nat. Acad. Sci. U.S.A.*, **35**, 1949, 94–99). With a few simplifications and addenda we have essentially followed Weil's presentation.

Exercises

1 Let p be a prime and $d = (m, p - 1)$. Prove that $N(x^m = a) = \sum \chi(a)$, the sum being over all χ such that $\chi^d = \varepsilon$.

2 With the notation of Exercise 1 show that $N(x^m = a) = N(x^d = a)$ and conclude that if $d_i = (m_i, p - 1)$, then $\sum_i a_i x^{m_i} = b$ and $\sum_i a_i x^{d_i} = b$ have the same number of solutions.

3 Let χ be a nontrivial multiplicative character of F_p and ρ be the character of order 2. Show that $\sum_t \chi(1 - t^2) = J(\chi, \rho)$. [*Hint:* Evaluate $J(\chi, \rho)$ using the relation $N(x^2 = a) = 1 + \rho(a)$.]

4 Show, if $p \nmid k$, that $\sum_t \chi(t(k - t)) = \chi(k^2/2^2)J(\chi, \rho)$. If χ is a character of odd order, show that the relationship remains true even if $p \mid k$.

5 If χ is a character of odd order, show that $g(\chi)^2 = \chi(2)^{-2}J(\chi, \rho)g(\chi^2)$. [*Hint:* Write out $g(\chi)^2$ explicitly and use Exercise 4.]

6 (continuation) If χ has odd order, show that $J(\chi, \chi) = \chi(2)^{-2}J(\chi, \rho)$.

7 Suppose that $p \equiv 1$ (4) and that χ is a character of order 4. Then $\chi^2 = \rho$ and $J(\chi, \chi) = \chi(-1)J(\chi, \rho)$. [*Hint:* Evaluate $g(\chi)^4$ in two ways.]

8 Generalize Exercise 3 in the following way. Suppose that p is a prime, $\sum_t \chi(1 - t^m) = \sum_\lambda J(\chi, \lambda)$, where λ varies over all characters such that $\lambda^m = \varepsilon$. Conclude that $|\sum_t \chi(1 - t^m)| \le (m - 1)p^{1/2}$.

9 Suppose that $p \equiv 1$ (3) and that χ is a character of order 3. Prove (using Exercise 5) that $g(\chi)^3 = p\pi$, where $\pi = \chi(2)J(\chi, \rho)$.

10 (continuation) Show that $\chi\rho$ is a character of order 6 and that $g(\chi\rho)^6 = (-1)^{(p-1)/2}p\bar{\pi}^4$.

11 Use Gauss's theorem to find the number of solutions to $x^3 + y^3 = 1$ in F for $p = 13, 19, 37$, and 97.

12 If $p \equiv 1$ (4), then we have seen that $p = a^2 + b^2$ with $a, b \in \mathbb{Z}$. If we require that a and b be positive, that a be odd, and that b be even, show that a and b are uniquely determined. (*Hint:* Use the fact that unique factorization holds in $\mathbb{Z}[i]$ and that if $p = a^2 + b^2$ then $a + bi$ is a prime in $\mathbb{Z}[i]$.)

13 If $p \equiv 1$ (3), we have seen that $4p = A^2 + 27B^2$ with $A, B \in \mathbb{Z}$. If we require that $A \equiv 1$ (3), show that A is uniquely determined. (*Hint:* Use the fact that unique factorization holds in $\mathbb{Z}[\omega]$. This proof is a little trickier than that for Exercise 12.)

14 Suppose that $p \equiv 1$ (n) and that χ is a character of order n. Show that $g(\chi)^n \in \mathbb{Z}[\zeta]$, where $\zeta = e^{2\pi i/n}$.

15 Suppose that $p \equiv 1$ (6) and let χ and ρ be characters of order 3 and 2, respectively. Show that the number of solutions to $y^2 = x^3 + D$ in F is $p + \pi + \bar{\pi}$, where $\pi = \chi\rho(D)J(\chi, \rho)$. If $\chi(2) = 1$, show that the number of solutions to $y^2 = x^3 + 1$ is $p + A$, where $4p = A^2 + 27B^2$ and $A \equiv 1$ (3). Verify this result numerically when $p = 31$.

16 Suppose that $p \equiv 1$ (4) and that χ is a character of order 4. Let N be the number of solutions to $x^4 + y^4 = 1$ in F_p. Show that $N = p + 1 - \delta_4(-1)4 + 2\operatorname{Re} J(\chi, \chi) + 4\operatorname{Re} J(\chi, \rho)$.

17 (continuation) By Exercise 7, $J(\chi, \chi) = \chi(-1)J(\chi, \rho)$. Let $\pi = -J(\chi, \rho)$. Show that
(a) $N = p - 3 - 6\operatorname{Re}\pi$ if $p \equiv 1\,(8)$.
(b) $N = p + 1 - 2\operatorname{Re}\pi$ if $p \equiv 5\,(8)$.

18 (continuation) Let $\pi = a + bi$. One can show (see Chapter 11, Section 5) that a is odd, b is even, and $a \equiv 1\,(4)$ if $4|b$ and $a \equiv -1\,(4)$ if $4 \nmid b$. Let $p = A^2 + B^2$ and fix A by requiring that $A \equiv 1$ (4). Then show that
(a) $N = p - 3 - 6A$ if $p \equiv 1\,(8)$.
(b) $N = p + 1 + 2A$ if $p \equiv 5\,(8)$.

19 Find a formula for the number of solutions to $x_1^2 + x_2^2 + \cdots + x_r^2 = 0$ in F_p.

20 Generalize Proposition 8.6.1 by finding an explicit formula for the number of solutions to $a_1x_1^2 + a_2x_2^2 + \cdots + a_rx_r^2 = 1$ in F_p.

21 Suppose that $p \equiv 1\,(d)$, $\zeta = e^{2\pi i/p}$, and consider $\sum_x \zeta^{ax^d}$. Show that $\sum_x \zeta^{ax^d} = \sum_r m(r)\zeta^{ar}$, where $m(r) = N(x^d = r)$.

22 (continuation) Prove that $\sum_x \zeta^{ax^d} = \sum_\chi g_a(\chi)$, where the sum is over all χ such that $\chi^d = \varepsilon$, $\chi \neq \varepsilon$. Assume that $p \nmid a$.

23 Let $f(x_1, x_2, \ldots, x_n) \in F_p[x_1, x_2, \ldots, x_n]$. Let N be the number of zeros of f in F_p. Show that $N = p^{n-1} + p^{-1}\sum_{a \neq 0}(\sum_{x_1,\ldots,x_n} \zeta^{af(x_1,\ldots,x_n)})$.

24 (continuation) Let $f(x_1, x_2, \ldots, x_n) = a_1x_1^{m_1} + a_2x_2^{m_2} + \cdots + a_nx_n^{m_n}$. Let $d_i = (m_i, p - 1)$. Show that $N = p^{n-1} + p^{-1}\sum_{a \neq 0}\prod_{i=1}^{n}\sum_{\chi_i} g_{aa_i}(\chi_i)$, where χ_i runs over all characters such that $\chi_i^{d_i} = \varepsilon$ and $\chi_i \neq \varepsilon$.

25 Deduce from Exercise 24 that $|N - p^{n-1}| \leq (p-1)(d_1 - 1)\cdots(d_n - 1)p^{(n/2)-1}$.

chapter nine/CUBIC RECIPROCITY

In Chapter 5 we saw that the law of quadratic reciprocity provided the answer to the question, For which primes p is the congruence $x^2 \equiv a\ (p)$ solvable? Here a is a fixed integer. If the same question is considered for congruences $x^n \equiv a\ (p)$, n a fixed positive integer, we are led into the realm of the higher reciprocity laws. When $n = 3$ and 4 we speak of cubic and biquadratic reciprocity.

In the introduction to his famous pair of papers, "Theorie der biquadratischen Rests I, II" [34], Gauss claims that the theory of quadratic residues had been brought to such a state of perfection that nothing more could be wished. On the other hand, "The theory of cubic and biquadratic residues is by far more difficult." He had only been able to deal with certain special cases for which the proofs had been so difficult that he soon came to the realization that "... the previously accepted principles of arithmetic are in no way sufficient for the foundations of a general theory, that rather such a theory necessarily demands that to a certain extent the domain of higher arithmetic needs to be endlessly enlarged ...." In modern language, he is calling for the establishment of a theory of algebraic numbers. As a first step, because this is what is needed for discussing biquadratic residues, he investigated in detail the arithmetic of the ring $\mathbb{Z}[\sqrt{-1}]$, which we now refer to as the ring of Gaussian integers.

Curiously, although Gauss formulated and discovered the law of biquadratic reciprocity, he did not prove it completely. The first complete proofs of cubic and biquadratic reciprocity are due to G. Eisenstein and C. G. Jacobi.

In this chapter we shall formulate and prove the law of cubic reciprocity. In fact, we shall give two proofs. The first is due to Eisenstein. It uses only Gauss sums and is in every way similar to the proof of quadratic reciprocity given in Chapter 6. The second uses Jacobi sums and is analogous to the proof of quadratic reciprocity given in Chapter 8, Section 6.

I THE RING $\mathbb{Z}[\omega]$

Let $\omega = (-1 + \sqrt{-3})/2$. The ring $\mathbb{Z}[\omega]$ was defined and discussed in Chapter 1, Section 4. Its elements are complex numbers of the form $a + b\omega$, $a, b \in \mathbb{Z}$. If $\alpha = a + b\omega \in \mathbb{Z}[\omega]$, define the norm of α, $N\alpha$, by the formula $N\alpha = \alpha\bar{\alpha} = a^2 - ab + b^2$. Here $\bar{\alpha}$ means the complex conjugate of α. In Chapter 1 we used the notation $\lambda(\alpha)$ instead of $N\alpha$. The change is merely a matter of conforming to standard notation. For notational convenience we shall set $D = \mathbb{Z}[\omega]$.

We have proved earlier that D is a unique factorization domain. Our first task here is to discover the units and the prime elements in D.

Proposition 9.1.1

$\alpha \in D$ is a unit iff $N\alpha = 1$. The only units in D are $1, -1, \omega, -\omega, \omega^2$, and $-\omega^2$.

PROOF

If $N\alpha = 1$, $\alpha\bar{\alpha} = 1$, which implies that α is a unit since $\bar{\alpha} \in D$.

If α is a unit, there is a $\beta \in D$ such that $\alpha\beta = 1$. Thus $N\alpha N\beta = 1$. Since $N\alpha$ and $N\beta$ are positive integers this implies that $N\alpha = 1$.

Now suppose that $\alpha = a + b\omega$ is a unit. Then $1 = a^2 - ab + b^2$ or $4 = (2a - b)^2 + 3b^2$. There are two possibilities:

(a) $2a - b = \pm 1, b = \pm 1$.
(b) $2a - b = \pm 2, b = 0$.

Solving these six pairs of equations yields the result $1, -1, \omega, -\omega, -1, -\omega$, and $1 + \omega$. Since $\omega^2 + \omega + 1 = 0$ the last two elements are ω^2 and $-\omega^2$. We are done.

To investigate primes in D it is important to realize that primes in $\mathbb{Z}$ need not be prime in D. For example, $7 = (3 + \omega)(2 - \omega)$. For this reason we shall speak of primes in $\mathbb{Z}$ as rational primes and refer to primes in D simply as primes.

Proposition 9.1.2

If π is a prime in D, then there is a rational prime p such that $N\pi = p$ or p^2. In the former case π is not associate to a rational prime; in the latter case π is associate to p.

PROOF

We have $N\pi = n > 1$, or $\pi\bar{\pi} = n$. n is a product of rational primes. Thus $\pi \mid p$ for some rational prime p. If $p = \pi\gamma$, $\gamma \in D$, then $N\pi N\gamma = Np = p^2$. Thus either $N\pi = p^2$ and $N\gamma = 1$ or $N\pi = p$. In the former case γ is a unit and therefore π is associate to p. In the latter case if $\pi = uq$, u a unit and q a rational prime, then $p = N\pi = NuNq = q^2$, which is nonsense. Thus π is not associate to a rational prime.

Proposition 9.1.3

If $\pi \in D$ is such that $N\pi = p$, a rational prime, then π is a prime in D.

PROOF

If π were not prime in D, then we could write $\pi = \rho\gamma$ with $N\rho, N\pi > 1$. Then $p = N\pi = N\rho N\gamma$, which cannot be true since p is prime in $\mathbb{Z}$. Thus π is a prime in D.

The following result classifies primes in D.

Proposition 9.1.4

Suppose that p and q are rational primes. If $q \equiv 2\,(3)$, then q is prime in D. If $p \equiv 1\,(3)$, then $p = \pi\bar{\pi}$, where π is prime in D. Finally $3 = -\omega^2(1-\omega)^2$, and $1-\omega$ is prime in D.

PROOF

Suppose that p were not a prime. Then $p = \pi\gamma$, with $N\pi > 1$, $N\gamma > 1$. Thus $p^2 = N\pi N\gamma$ and $N\pi = p$. Let $\pi = a + b\omega$. Then $p = a^2 - ab + b^2$ or $4p = (2a-b)^2 + 3b^2$, yielding $p \equiv (2a-b)^2\,(3)$. If $3 \nmid p$ we have $p \equiv 1\,(3)$ for 1 is the only nonzero square mod 3. It follows immediately that if $q \equiv 2\,(3)$, it is a prime in D.

Now, suppose that $p \equiv 1\,(3)$. By quadratic reciprocity we have

$$\begin{aligned}(-3/p) &= (-1/p)(3/p) = (-1)^{(p-1)/2}(p/3)(-1)^{((p-1)/2)((3-1)/2)}\\ &= (p/3) = (1/3) = 1\end{aligned}$$

Hence, there is an $a \in \mathbb{Z}$ such that $a^2 \equiv -3\,(p)$ or $pb = a^2 + 3$ for some $b \in \mathbb{Z}$. Thus p divides $(a + \sqrt{-3})\,(a - \sqrt{-3}) = (a + 1 + 2\omega) \times (a - 1 + 2\omega)$. If p were a prime in D, it would have to divide one of the factors but this cannot happen since $p \neq 2$ and $2/p \notin \mathbb{Z}$. Thus $p = \pi\gamma$ with π and γ nonunits. Taking norms we see that $p^2 = N\pi N\gamma$ and that $p = N\pi = \pi\bar{\pi}$.

The last case is handled as follows. $x^3 - 1 = (x-1)(x-\omega) \times (x-\omega^2)$ implies that $x^2 + x + 1 = (x-\omega)(x-\omega^2)$. Setting $x = 1$ yields $3 = (1-\omega)(1-\omega^2) = (1+\omega)(1-\omega)^2 = -\omega^2(1-\omega)^2$. Taking norms we see that $9 = N(1-\omega)^2$ and so $3 = N(1-\omega)$. Thus $1-\omega$ is a prime.

As a matter of notation q will be a rational prime congruent to 2 modulo 3 and π a complex prime whose norm, $N\pi = p$, is a rational prime congruent to 1 modulo 3. Occasionally π will refer to an arbitrary prime of D. The context should make the usage clear.

2 RESIDUE CLASS RINGS

Just as in the ring $\mathbb{Z}$ and in the ring of all algebraic integers, the notion of congruence is extremely useful in D. If α, β, $\gamma \in D$ and $\gamma \neq 0$ is a nonunit, we say that $\alpha \equiv \beta\ (\gamma)$ if γ divides $\alpha - \beta$. Just as in $\mathbb{Z}$ the congruence classes modulo γ may be made into a ring $D/\gamma D$, called the residue class ring modulo γ.

Proposition 9.2.1

Let $\pi \in D$ be a prime. Then $D/\pi D$ is a finite field with $N\pi$ elements.

PROOF

We first show that $D/\pi D$ is a field. Let $\alpha \in D$ be such that $\alpha \not\equiv 0\ (\pi)$. By Corollary 1 to Proposition 1.3.2 there exist elements $\beta, \gamma \in D$ such that $\beta\alpha + \gamma\pi = 1$. Thus $\beta\alpha \equiv 1\ (\pi)$, which shows that the residue class of α is a unit in $D/\pi D$.

To show that $D/\pi D$ has $N\pi$ elements we must consider separately the cases in Proposition 9.1.3.

Suppose that $\pi = q$ is a rational prime congruent to 2 modulo 3. We claim that $\{a + b\omega \mid 0 \leq a < q \text{ and } 0 \leq b < q\}$ is a complete set of coset representatives. This will show that D/qD has $q^2 = Nq$ elements. Let $\mu = m + n\omega \in D$. Then $m = qs + a$ and $n = qt + b$, where s, t, a, $b \in \mathbb{Z}$ and $0 \leq a$, $b < q$. Clearly $\mu \equiv a + b\omega\ (q)$. Next, suppose that $a + b\omega \equiv a' + b'\omega\ (q)$, where $0 \leq a, b, a', b' < q$. Then $((a - a')/q) + ((b - b')/q)\omega \in D$, implying that $(a - a')/q$ and $(b - b')/q$ are in $\mathbb{Z}$. This is possible only if $a = a'$ and $b = b'$.

Now suppose that $p \equiv 1\ (3)$ is a rational prime and $\pi\bar{\pi} = N\pi = p$. We claim that $\{0, 1, \ldots, p - 1\}$ is a complete set of coset representatives. This will show that $D/\pi D$ has $p = N\pi$ elements. Let $\pi = a + b\omega$. Since $p = a^2 - ab + b^2$ it follows that $p \nmid b$. Let $\mu = m + n\omega$. There is an integer c such that $cb \equiv n\ (p)$. Then $\mu - c\pi \equiv m - ca\ (p)$ and so $\mu \equiv m - ca\ (\pi)$. Every element of D is congruent to a rational integer modulo π. If $l \in \mathbb{Z}$, $l = sp + r$, where $s, r \in \mathbb{Z}$ and $0 \leq r < p$. Thus $l \equiv r\ (p)$ and a fortiori $l \equiv r(\pi)$. We have shown that every element of D is congruent to an element of $\{0, 1, 2, \ldots, p - 1\}$ modulo π. If $r \equiv r'\ (\pi)$ with $r, r' \in \mathbb{Z}$ and $0 \leq r, r' < p$, then $r - r' = \pi\gamma$ and $(r - r')^2 = pN\gamma$, implying that $p \mid r - r'$. Thus $r = r'$ and we are done.

We leave the case of the prime $1 - \omega$ as an exercise.

3 CUBIC RESIDUE CHARACTER

Let π be a prime. Then the multiplicative group of $D/\pi D$ has order $N\pi - 1$. Hence we have an analog of Fermat's little theorem.

Proposition 9.3.1
If $\pi \nmid \alpha$, then

$$\alpha^{N\pi - 1} \equiv 1 \ (\pi)$$

If the norm of π is different from 3, then the residue classes of 1, ω, and ω^2 are distinct in $D/\pi D$. To see this, suppose, for example, that $\omega \equiv 1$ (π). Then $\pi \mid (1 - \omega)$, and since $1 - \omega$ is prime, π and $1 - \omega$ are associate. Thus $N\pi = N(1 - \omega) = 3$, a contradiction. The other cases are handled in the same way.

Since $\{1, \omega, \omega^2\}$ is a cyclic group of order 3 it follows that 3 divides the order of $(D/\pi D)^*$; i.e., $3 \mid N\pi - 1$. This can be seen in another way using Proposition 9.1.3. If $\pi = q$, a rational prime, then $N\pi = q^2 \equiv 1$ (3). If π is such that $N\pi = p$, then $p \equiv 1$ (3).

Proposition 9.3.2
Suppose that π is a prime such that $N\pi \neq 3$ and that $\pi \nmid \alpha$. Then there is a unique integer $m = 0, 1,$ or 2 such that $\alpha^{(N\pi - 1)/3} \equiv \omega^m$ (π).

PROOF
We know that π divides $\alpha^{N\pi - 1} - 1$. Now,

$$\alpha^{N\pi - 1} - 1 = (\alpha^{(N\pi - 1)/3} - 1)(\alpha^{(N\pi - 1)/3} - \omega)(\alpha^{(N\pi - 1)/3} - \omega^2)$$

Since π is prime it must divide one of the three factors on the right. By the preceding remarks it can divide at most one factor, since if it divided two factors it would divide the difference. This proves the proposition.

On the basis of this result we can make the following definition.

Definition
If $N\pi \neq 3$, the *cubic residue character* of α modulo π is given by

(a) $(\alpha/\pi)_3 = 0$ if $\pi \mid \alpha$.

(b) $\alpha^{(N\pi - 1)/3} \equiv (\alpha/\pi)_3$ (π), with $(\alpha/\pi)_3$ equal to 1, ω, or ω^2.

This character plays the same role in the theory of cubic residues as the Legendre symbol plays in the theory of quadratic residues.

Proposition 9.3.3

(a) $(\alpha/\pi)_3 = 1$ *iff* $x^3 \equiv \alpha$ (π) *is solvable, i.e., iff* α *is a cubic residue.*

(b) $\alpha^{(N\pi - 1)/3} \equiv (\alpha/\pi)_3$ (π).

(c) $(\alpha\beta/\pi)_3 = (\alpha/\pi)_3(\beta/\pi)_3$.
(d) *If* $\alpha \equiv \beta\ (\pi)$, *then* $(\alpha/\pi)_3 = (\beta/\pi)_3$.

PROOF

Part (a) is a special case of Proposition 7.1.2. Take $F = D/\pi D$, $q = N\pi$, and $n = 3$ in that proposition.

Part (b) is immediate from the definition.

Part (c): $(\alpha\beta/\pi)_3 \equiv (\alpha\beta)^{(N\pi-1)/3} \equiv \alpha^{(N\pi-1)/3}\beta^{(N\pi-1)/3} \equiv (\alpha/\pi)_3(\beta/\pi)_3$. The result follows.

Part (d): If $\alpha \equiv \beta\ (\pi)$, then $(\alpha/\pi)_3 \equiv \alpha^{(N\pi-1)/3} \equiv \beta^{(N\pi-1)/3} \equiv (\beta/\pi)_3\ (\pi)$, and so $(\alpha/\pi)_3 = (\beta/\pi)_3$.

Since we shall be dealing only with cubic characters in this chapter the notation $\chi_\pi(\alpha) = (\alpha/\pi)_3$ will be convenient.

It is useful to study the behavior of characters under complex conjugation.

Proposition 9.3.4

(a) $\overline{\chi_\pi(\alpha)} = \chi_\pi(\alpha)^2 = \chi_\pi(\alpha^2)$.
(b) $\overline{\chi_\pi(\alpha)} = \chi_{\bar\pi}(\bar\alpha)$.

PROOF

(a) $\chi_\pi(\alpha)$ is by definition 1, ω, or ω^2, and each of these numbers squared is equal to its conjugate.

(b) From

$$\alpha^{(N\pi-1)/3} \equiv \chi_\pi(\alpha)\ (\pi)$$

we get

$$\bar\alpha^{(N\pi-1)/3} \equiv \overline{\chi_\pi(\alpha)}\ (\bar\pi)$$

Since $N\bar\pi = N\pi$ this shows that $\chi_{\bar\pi}(\bar\alpha) \equiv \overline{\chi_\pi(\alpha)}\ (\pi)$ and thus that $\chi_{\bar\pi}(\bar\alpha) = \overline{\chi_\pi(\alpha)}$.

Corollary

$\chi_q(\bar\alpha) = \chi_q(\alpha^2)$ *and* $\chi_q(n) = 1$ *if n is a rational integer prime to q.*

PROOF

Since $\bar q = q$ we have $\chi_q(\bar\alpha) = \chi_{\bar q}(\bar\alpha) = \overline{\chi_q(\alpha)} = \chi_q(\alpha^2)$. This gives the first relation.

Since $\bar n = n$ we have $\chi_q(n) = \chi_q(n^2) = \chi_q(n)^2$. Since $\chi_q(n) \neq 0$ it follows that $\chi_q(n) = 1$.

The corollary states that n is a cubic residue modulo q. Thus, if $q_1 \neq q_2$ are two primes congruent to 2 modulo 3, then we have (trivially) $\chi_{q_1}(q_2) = \chi_{q_2}(q_1)$. This is a special case of the law of cubic reciprocity. To formulate the general law we need to introduce the idea of a "primary" prime.

Definition

If π is a prime in D, we say that π is *primary* if $\pi \equiv 2\,(3)$.

If $\pi = q$ is rational, this is nothing new. If $\pi = a + b\omega$ is a complex prime, the definition is equivalent to $a \equiv 2\,(3)$ and $b \equiv 0\,(3)$.

We need a notion such as "primary" to eliminate the ambiguity caused by the fact that every nonzero element of D has six associates.

Proposition 9.3.5

Suppose that $N\pi = p \equiv 1\,(3)$. *Among the associates of* π *exactly one is primary.*

PROOF

Write $\pi = a + b\omega$. The associates of π are π, $\omega\pi$, $\omega^2\pi$, $-\pi$, $-\omega\pi$, and $-\omega^2\pi$. In terms of a and b these elements can be expressed as

(a) $a + b\omega$.

(b) $-b + (a - b)\omega$.

(c) $(b - a) - a\omega$.

(d) $-a - b\omega$.

(e) $b + (b - a)\omega$.

(f) $(a - b) + a\omega$.

Since $p = a^2 - ab + b^2$, not both a and b are divisible by 3. By looking at parts (a) and (b) it is clear that we can assume that $3 \nmid a$. Considering parts (a) and (d) we can assume further that $a \equiv 2\ (3)$. Under this assumption $p = a^2 - ab + b^2$ leads to $1 \equiv 4 - 2b + b^2\ (3)$ or $b(b - 2) \equiv 0\ (3)$. If $3 \mid b$, then $a + b\omega$ is primary. If $b \equiv 2\ (3)$, then $b + (b - a)\omega$ is primary.

To show uniqueness, assume that $a + b\omega$ is primary. By considering the congruence class of the first term in part (b) to part (e) we see that none of these expressions is primary. Neither is the expression in part (f) since the coefficient of ω, a, is not divisible by 3.

For example, $3 + \omega$ is prime since $N(3 + \omega) = 7$, and $-\omega^2(3 + \omega) = 2 + 3\omega$ is the primary prime associated to it.

We can now state

Theorem 1 (The Law of Cubic Reciprocity)
Let π_1 and π_2 be primary, $N\pi_1, N\pi_2 \neq 3$, and $N\pi_1 \neq N\pi_2$. Then

$$\chi_{\pi_1}(\pi_2) = \chi_{\pi_2}(\pi_1)$$

A proof will be given in Section 4, but first a few remarks are in order.

(a) There are three cases to consider. Namely, both π_1 and π_2 are rational, π_1 is rational and π_2 is complex, and both π_1 and π_2 are complex. The first case is, as we have seen, trivial.

(b) The cubic character of the units can be dealt with as follows. Since $-1 = (-1)^3$ we have $\chi_\pi(-1) = 1$ for all primes π.

If $N\pi \neq 3$, then it follows from Proposition 9.3.3, part (b), that $\chi_\pi(\omega) = \omega^{(N\pi-1)/3}$. Thus $\chi_\pi(\omega) = 1, \omega$, or ω^2 according to whether $N\pi = 1, 4$, or 7 modulo 9.

(c) The prime $1 - \omega$ causes particular difficulty. If $N\pi \neq 3$, we would like to evaluate $\chi_\pi(1 - \omega)$. This is done by Eisenstein in [29] by a highly ingenious argument. We shall content ourselves with stating the result and proving a special case.

Theorem 1′ (Supplement to the Cubic Reciprocity Law)
Suppose that $N\pi \neq 3$. If $\pi = q$ is rational, write $q = 3m - 1$. If $\pi = a + b\omega$ is a primary complex prime, write $a = 3m - 1$. Then

$$\chi_\pi(1 - \omega) = \omega^{2m}$$

We give a proof for the case of a rational prime q. Since $(1 - \omega)^2 = -3\omega$ we have

$$\chi_q(1 - \omega)^2 = \chi_q(-3)\chi_q(\omega)$$

By the corollary to Proposition 9.3.4 we know that $\chi_q(-3) = 1$. By remark (b) $\chi_q(\omega) = \omega^{(Nq-1)/3} = \omega^{(q^2-1)/3}$. Thus $\chi_q(1 - \omega)^2 = \omega^{(q^2-1)/3}$. Squaring both sides yields

$$\chi_q(1 - \omega) = \omega^{(2/3)(q^2-1)}$$

Now, $q^2 - 1 = 9m^2 - 6m$ so that $\frac{2}{3}(q^2 - 1) \equiv -4m \equiv 2m\,(3)$. The result follows.

4 PROOF OF THE LAW OF CUBIC RECIPROCITY

Let π be a complex prime such that $N\pi = p \equiv 1\,(3)$. Since $D/\pi D$ is a finite field of characteristic p it contains a copy of $\mathbb{Z}/p\mathbb{Z}$. Both $D/\pi D$ and $\mathbb{Z}/p\mathbb{Z}$ have p elements. Thus we may identify the two fields. More explicitly the identification is given by sending the coset of n in Z/pZ to the coset of n in $D/\pi D$.

This identification allows us to consider χ_π as a cubic character on Z/pZ in the sense of Chapter 8 [see Proposition 9.3.3, parts (c) and (d)]. Thus we may work with the Gauss sums $g_a(\chi_\pi)$ and the Jacobi sum $J(\chi_\pi, \chi_\pi)$.

If χ is any cubic character, we have proved (see the corollaries to Proposition 8.3.3 and Proposition 8.3.4) that

(a) $g(\chi)^3 = pJ(\chi, \chi)$.

(*b*) If $J(\chi, \chi) = a + b\omega$, then $a \equiv -1\ (3)$ and $b \equiv 0\ (3)$.

Since $J(\chi, \chi)\overline{J(\chi, \chi)} = p$, the second assertion says that $J(\chi, \chi)$ is a primary prime in D of norm p.

We need a lemma.

Lemma 1

$J(\chi_\pi, \chi_\pi) = \pi$.

PROOF

Let $J(\chi_\pi, \chi_\pi) = \pi'$. Since $\pi\bar{\pi} = p = \pi'\bar{\pi}'$ we have $\pi \mid \pi'$ or $\pi \mid \bar{\pi}'$.

Since all the primes involved are primary we must have $\pi = \pi'$ or $\pi = \bar{\pi}'$. We wish to eliminate the latter possibility.

From the definitions,

$$J(\chi_\pi, \chi_\pi) = \sum_x \chi_\pi(x)\chi_\pi(1 - x) \equiv \sum_x x^{(p-1)/3}(1 - x)^{(p-1)/3}\ (\pi)$$

where the sum is over $\mathbb{Z}/p\mathbb{Z}$. The polynomial $x^{(p-1)/3}(1 - x)^{(p-1)/3}$ is of degree $\frac{2}{3}(p - 1) < p - 1$. By Exercise 11 of Chapter 4 it follows that $\sum_x x^{(p-1)/3}(1 - x)^{(p-1)/3} \equiv 0\ (p)$. This shows that $J(\chi_\pi, \chi_\pi) \equiv 0\ (\pi)$; i.e., $\pi \mid \pi'$ and therefore $\pi = \pi'$.

Corollary

$g(\chi_\pi)^3 = p\pi$.

We can now prove the law of cubic reciprocity. We first consider the case where $\pi_1 = q \equiv 2\ (3)$ and $\pi_2 = \pi$ with $N\pi = p$.

Raise both sides of the relation $g(\chi_\pi)^3 = p\pi$ to the $(q^2 - 1)/3$ power. This gives $g(\chi_\pi)^{q^2-1} = (p\pi)^{(q^2-1)/3}$. Taking congruences modulo q we see that

$$g(\chi_\pi)^{q^2-1} \equiv \chi_q(p\pi)\ (q)$$

Since $\chi_q(p) = 1$ this leads to

$$g(\chi_\pi)^{q^2} \equiv \chi_q(\pi)g(\chi_\pi)\ (q) \tag{1}$$

We now analyze the left-hand side:

$$g(\chi_\pi)^{q^2} = \left(\sum \chi_\pi(t)\xi^t\right)^{q^2} \equiv \sum \chi_\pi(t)^{q^2}\xi^{q^2t}\ (q)$$

Since $q^2 \equiv 1\ (3)$ and $\chi_\pi(t)$ is a cube root of 1 we have

$$g(\chi_\pi)^{q^2} \equiv g_{q^2}(\chi_\pi)\ (q) \tag{2}$$

By Proposition 8.2.1 $g_{q^2}(\chi_\pi) = \chi_\pi(q^{-2})g(\chi_\pi) = \chi_\pi(q)g(\chi_\pi)$. Thus, combining Equations (1) and (2)

$$\chi_\pi(q)g(\chi_\pi) \equiv \chi_q(\pi)g(\chi_\pi)\ (q)$$

Multiply both sides of this congruence by $\overline{g(\chi_\pi)}$. Since $g(\chi_\pi)\overline{g(\chi_\pi)} = p$,

$$\chi_\pi(q)p \equiv \chi_q(\pi)p\ (q)$$

or

$$\chi_\pi(q) \equiv \chi_q(\pi)\ (q)$$

implying that

$$\chi_\pi(q) = \chi_q(\pi)$$

It remains to consider the case of two complex primes π_1 and π_2, where $N\pi_1 = p_1 \equiv 1\ (3)$ and $N\pi_2 = p_2 \equiv 1\ (3)$. This case is handled by essentially the same technique, but it is a little trickier.

Let $\gamma_1 = \bar{\pi}_1$ and $\gamma_2 = \bar{\pi}_2$. Then γ_1 and γ_2 are primary and $p_1 = \pi_1\gamma_1$ and $p_2 = \pi_2\gamma_2$.

Starting from the relation $g(\chi_{\gamma_1})^3 = p_1\gamma_1$, raising to the $(N\pi_2 - 1)/3 = (p_2 - 1)/3$ power, and taking congruences modulo π_2, we obtain by the same method as above the relation

$$\chi_{\gamma_1}(p_2^2) = \chi_{\pi_2}(p_1\pi_1) \tag{3}$$

Similarly, starting from $g(\chi_{\pi_2})^3 = p_2\pi_2$, raising to the $(p_1 - 1)/3$ power, and taking congruences modulo π_1, we obtain

$$\chi_{\pi_2}(p_1^2) = \chi_{\pi_1}(p_2\pi_2) \tag{4}$$

We also need the relation $\chi_{\gamma_1}(p_2^2) = \chi_{\pi_1}(p_2)$, which follows from Proposition 9.3.4 since $\gamma_1 = \bar{\pi}_1$ and $\bar{p}_2 = p_2$. Now we calculate

$$\begin{aligned}
\chi_{\pi_1}(\pi_2)\chi_{\pi_2}(p_1\gamma_1) &= \chi_{\pi_1}(\pi_2)\chi_{\gamma_1}(p_2^2) && \text{by Equation (3)} \\
&= \chi_{\pi_1}(\pi_2)\chi_{\pi_1}(p_2) = \chi_{\pi_1}(p_2\pi_2) && \text{by above remark} \\
&= \chi_{\pi_2}(p_1^2) = \chi_{\pi_2}(p_1\pi_1\gamma_1) && \text{by Equation (4)} \\
&= \chi_{\pi_2}(\pi_1)\chi_{\pi_2}(p_1\gamma_1)
\end{aligned}$$

Equating the first and last terms and canceling $\chi_{\pi_2}(p_1\gamma_1)$ gives the sought for result:

$$\chi_{\pi_1}(\pi_2) = \chi_{\pi_2}(\pi_1)$$

5 ANOTHER PROOF OF THE LAW OF CUBIC RECIPROCITY

We present a proof of cubic reciprocity using Jacobi sums. This proof is somewhat shorter and more elegant than the one given in Section 4. It should be noticed, however, that more background material is used.

Consider the case $\pi_1 = q$, $\pi_2 = \pi$. Let $\chi_\pi = \chi$, and consider the Jacobi sum $J(\chi, \chi, \ldots, \chi)$ with q terms. Since $3 \mid q + 1$ we have by Corollary 1 to Theorem 3 of Chapter 8,

$$g(\chi)^{q+1} = qJ(\chi, \chi, \ldots, \chi) \tag{5}$$

Since $g(\chi)^3 = p\pi$,

$$g(\chi)^{q+1} = (p\pi)^{(q+1)/3} \tag{6}$$

Now, recall that

$$J(\chi, \chi, \ldots, \chi) = \sum \chi(x_1)\chi(x_2)\cdots\chi(x_q)$$

where the sum is over all $x_1, x_2, \ldots, x_q \in Z/pZ$ such that $x_1 + x_2 + \cdots + x_q = 1$. Consider the term for which $x_1 = x_2 = \cdots = x_q$. Then $qx_1 = 1$ and $\chi(q)\chi(x_1) = 1$. Raising both sides to the qth power, and recalling that $q \equiv 2\ (3)$, yields $\chi(q)^2\chi(x_1)^q = 1$ and so $\chi(x_1)^q = \chi(q)$. Thus the "diagonal" term of $J(\chi, \chi, \ldots, \chi)$ has the value $\chi(q)$. If not all the x_1 are equal, there are q different q-tuples obtained from $(x_1, x_2, \ldots, x_q)$ by cyclic permutation. The corresponding terms of $J(\chi, \chi, \ldots, \chi)$ all have the same value. Thus

$$J(\chi, \chi, \ldots, \chi) \equiv \chi(q)\ (q) \tag{7}$$

Combining Equations (5), (6), and (7) we obtain

$$(p\pi)^{(q+1)/3} \equiv p\chi(q)\ (q)$$

or

$$p^{(q-2)/3}\pi^{(q+1)/3} \equiv \chi(q)\ (q)$$

Raising both sides to the $q - 1$ power (remember that $q - 1 \equiv 1\ (3)$)

$$p^{((q-2)/3)(q-1)}\pi^{(q^2-1)/3} \equiv \chi(q)^{q-1} \equiv \chi(q)(q)$$

Since $p^{((q-2)/3)(q-1)} \equiv 1$ (q) by Fermat's theorem and $\pi^{(q^2-1)/3} \equiv \chi_q(\pi)$ (q) it follows that

$$\chi_q(\pi) \equiv \chi_\pi(q)\ (q)$$

and

$$\chi_q(\pi) = \chi_\pi(q)$$

Now consider the case of two primary complex primes π_1 and π_2. Let $\gamma_1 = \bar{\pi}_1, \gamma_2 = \bar{\pi}_2, p_1 = \pi_1\gamma_1$, and $p_2 = \pi_2\gamma_2$. Then $p_1, p_2 \equiv 1$ (3). By Theorem 3 of Chapter 8 we have

$$g(\chi_{\gamma_1})^{p_2} = J(\chi_{\gamma_1}, \ldots, \chi_{\gamma_1})g(\chi_{\gamma_1}^{p_2})$$

There are p_2 terms in the Jacobi sum. Since $p_2 \equiv 1$ (3), $\chi_{\gamma_1}^{p_2} = \chi_{\gamma_1}$. Thus

$$[g(\chi_{\gamma_1})^3]^{(p_2-1)/3} = J(\chi_{\gamma_1}, \ldots, \chi_{\gamma_1}) \tag{8}$$

By isolating the diagonal term of the Jacobi sum (as we have done a number of times by now) we find that

$$J(\chi_{\gamma_1}, \ldots, \chi_{\gamma_1}) \equiv \chi_{\gamma_1}(p_2^{-1}) \equiv \chi_{\gamma_1}(p_2^2)\ (p_2)$$

Using this and the fact that $g(\chi_{\gamma_1})^3 = p_1\gamma_1$, we obtain from Equation (8) the congruence

$$\chi_{\pi_2}(p_1\gamma_1) \equiv \chi_{\gamma_1}(p_2^2)\ (\pi_2)$$

and therefore

$$\chi_{\pi_2}(p_1\gamma_1) = \chi_{\gamma_1}(p_2^2) \tag{9}$$

Similarly one proves that

$$\chi_{\pi_1}(p_2\pi_2) = \chi_{\pi_2}(p_1^2) \tag{10}$$

Equations (9) and (10) are the basic relations. From here on one proceeds exactly as in Section 4 to the desired conclusion $\chi_{\pi_1}(\pi_2) = \chi_{\pi_2}(\pi_1)$.

6 THE CUBIC CHARACTER OF 2

The law of cubic reciprocity can be used to develop the theory of cubic residues in the same manner as the law of quadratic reciprocity led to the results of Chapter 5, Section 2. We shall forego a development of the general theory in favor of a discussion of an illuminating special case. Namely, we shall ask for all primes π in D for which 2 is a cubic residue.

To begin with, notice that $x^3 \equiv 2\ (\pi)$ is solvable iff $x^3 \equiv 2\ (\pi')$ is solvable for any associate of π. Thus we may assume that π is primary. If $\pi = q$ is a rational prime, then $\chi_q(2) = 1$ and so 2 is a cubic residue for all such primes. We assume from now on that $\pi = a + b\omega$ is a primary complex prime. By cubic reciprocity $\chi_\pi(2) = \chi_2(\pi)$. The norm of 2 is $2^2 = 4$. Thus

$$\pi = \pi^{(4-1)/3} \equiv \chi_2(\pi)\,(2)$$

It follows that $\chi_\pi(2) = 1$ iff $\pi \equiv 1\ (2)$. We have proved

Proposition 9.6.1
$x^2 \equiv 2\,(\pi)$ is solvable iff $\pi \equiv 1\ (2)$, i.e., iff $a \equiv 1\ (2)$ and $b \equiv 0\ (2)$.

It is possible to formulate this proposition in another way. Let $\pi = a + b\omega$ be a primary complex prime and $p = N\pi = a^2 - ab + b^2$. Then $4p = (2a - b)^2 + 3b^2$. If we set $A = 2a - b$ and $B = b/3$, then $4p = A^2 + 27B^2$. According to Proposition 8.3.2 the integers A and B are uniquely determined up to sign.

Proposition 9.6.2
If $p \equiv 1\ (3)$, then $x^3 \equiv 2\ (p)$ is solvable iff there are integers C and D such that $p = C^2 + 27D^2$.

PROOF
If $x^3 \equiv 2\,(p)$ is solvable, so is $x^3 \equiv 2\,(\pi)$ and thus $\pi \equiv 1\,(2)$ by Proposition 9.6.1. We have

$$4p = A^2 + 27B^2, \qquad \text{where } A = 2a - b,\ B = b/3$$

Since b is even, so are B and A. Let $D = B/2$ and $C = A/2$. Then $p = C^2 + 27D^2$.

Suppose, conversely, that $p = C^2 + 27D^2$. Then $4p = (2C)^2 + 27(2D)^2$. By uniqueness $B = \pm 2D$; i.e., B is even and thus so is b. It follows that $\pi = a + b\omega \equiv 1\ (2)$, and $x^3 \equiv 2\,(\pi)$ is solvable. Since $D/\pi D$ has $p = N\pi$ elements there is an integer h such that $h^3 \equiv 2\,(\pi)$. It is now easy to show that $h^3 \equiv 2\,(p)$. If $\pi \mid h^3 - 2$, then $\bar{\pi} \mid h^3 - 2$ and $\pi\bar{\pi} = p \mid (h^3 - 2)^2$. Consequently, $p \mid h^3 - 2$ and we are done.

As an example take $p = 7$. Then $x^3 \equiv 2\,(7)$ is not solvable since there are clearly no integers C and D such that $7 = C^2 + 27D^2$.

On the other hand, $p = 31 = 2^2 + 27 \cdot 1^2$. Thus $x^3 \equiv 2\,(31)$ is solvable. Indeed, $4^3 \equiv 2\,(31)$.

Notes
Gauss once remarked that Eisenstein, Archimedes, and Newton were the greatest mathematicians of all time. Without detracting from the

high merit of Eisenstein, most mathematicians find this judgment somewhat peculiar. Perhaps Gauss's praise is related to the fact that Eisenstein was the first to publish complete proofs of the laws of cubic and biquadratic reciprocity. These results appeared in 1844. Jacobi claimed to have proved the law of cubic reciprocity some years earlier. The dispute over priority appears to have been quite bitter. The interested reader can piece together the details from the following references: Cassels [15], Smith [72], Kummer [51], Eisenstein [30], and Jacobi [47].

We mentioned in the Notes to Chapter 6 Gauss's success in determining the sign of the Gauss sum attached to a quadratic character. The corresponding question for cubic Gauss sums is unsolved. To be more precise, notice that in the corollary to Lemma 1 the cube of the Gauss sum attached to the character χ_π is known precisely. Thus the Gauss sum itself is determined up to multiplication by 1, ω, or ω^2. The exact value has been discussed by Kummer, who made a conjecture about it that does not seem to account well for the numerical evidence that has been compiled with the use of computers. See the last section of Hasse's book [41] for a discussion. Recently, J. W. S. Cassels has published a paper [15] in which he makes an explicit conjecture that involves the values of certain elliptic functions. He includes supporting numerical evidence.

For an interesting discussion of the cubic, quartic, and quintic character of 2 and 3, see Lehmer ([55] and [56]).

Exercises

1 If $\alpha \in \mathbb{Z}[\omega]$, show that α is congruent to either 0, 1, or -1 modulo $1-\omega$.

2 From now on we shall set $D = \mathbb{Z}[\omega]$ and $\lambda = 1 - \omega$. For μ in D show that we can write $\mu = (-1)^a\omega^b\lambda^c\pi_1^{a_1}\pi_2^{a_2}\cdots\pi_t^{a_t}$, where a, b, c, and the a_i are nonnegative integers and the π_i are primary primes.

3 Let γ be a primary prime. To evaluate $\chi_\gamma(\mu)$ we see, by Exercise 2, that it is enough to evaluate $\chi_\gamma(-1)$, $\chi_\gamma(\omega)$, $\chi_\gamma(\lambda)$, and $\chi_\gamma(\pi)$, where π is a primary prime. Since $-1 = (-1)^3$ we have $\chi_\gamma(-1) = 1$. We now consider $\chi_\gamma(\omega)$. Let $\gamma = a + b\omega$ and set $a = 3m - 1$ and $b = 3n$. Show that $\chi_\gamma(\omega) = \omega^{m+n}$.

4 (continuation) Show that $\chi_\gamma(\omega) = 1$, ω, or ω^2 according to whether γ is congruent to 8, 2, or 5 modulo 3λ. In particular, if q is a rational prime, $q \equiv 2\ (3)$, then $\chi_q(\omega) = 1$, ω, or ω^2 according to whether $q \equiv 8, 2$, or $5\ (9)$. [*Hint:* $\gamma = a + b\omega = -1 + 3(m + n\omega)$, and so $\gamma \equiv -1 + 3(m+n)\ (3\lambda)$.]

5 In the text we stated Eisenstein's result $\chi_\gamma(\lambda) = \omega^{2m}$. Show that $\chi_\gamma(3) = \omega^{2n}$.

6 Prove that

(a) $\chi_\gamma(\lambda) = 1$ for $\gamma \equiv 8, 8 + 3\omega, 8 + 6\omega\ (9)$.

(b) $\chi_\gamma(\lambda) = \omega$ for $\gamma \equiv 5, 5 + 3\omega, 5 + 6\omega\ (9)$.

(c) $\chi_\gamma(\lambda) = \omega^2$ for $\gamma \equiv 2, 2 + 3\omega, 2 + 6\omega\ (9)$.

7 Find primary primes associate to $1 - 2\omega$, $-7 - 3\omega$, and $3 - \omega$.

8 Factor the following numbers into primes in D: 7, 21, 45, 22, and 143.

9 Show that $\bar{\alpha}$, the residue class of α, is a cube in the field $D/\pi D$ iff $\alpha^{(N\pi-1)/3} \equiv 1\ (\pi)$. Conclude that there are $(N\pi - 1)/3$ cubes in $D/\pi D$.

10 What is the factorization of $x^{24} - 1$ in $D/5D$?

11 How many cubes are there in $D/5D$?

12 Show that $\omega\lambda$ has order 8 in $D/5D$ and that $\omega^2\lambda$ has order 24. [*Hint:* Show first that $(\omega\lambda)^2$ has order 4.]

13 Show that π is a cube in $D/5D$ iff $\pi \equiv 1, 2, 3, 4, 1 + 2\omega, 2 + 4\omega, 3 + \omega$, or $4 + 3\omega\ (5)$.

14 For which primes $\pi \in D$ is $x^3 \equiv 5\ (\pi)$ solvable?

15 Suppose that $p \equiv 1\ (3)$ and that $p = \pi\bar{\pi}$, where π is a primary prime in D. Show that $x^3 \equiv a\ (p)$ is solvable in $\mathbb{Z}$ iff $\chi_\pi(a) = 1$. We assume that $a \in \mathbb{Z}$.

16 Is $x^3 \equiv 2 - 3\omega\ (11)$ solvable? Since $D/11D$ has 121 elements this is hard to resolve by straightforward checking. Fill in the details of the following proof that it is not solvable. $\chi_\pi(2 - 3\omega) = \chi_{2-3\omega}(11)$ and so we shall have a solution iff $x^3 \equiv 11(2 - 3\omega)$ is solvable. This congruence is solvable iff $x^3 \equiv 11\ (7)$ is solvable in $\mathbb{Z}$. However, $x^3 \equiv a\ (7)$ is solvable in $\mathbb{Z}$ iff $a \equiv 1$ or $6\ (7)$.

17 An element $\gamma \in D$ is called primary if $\gamma \equiv 2\ (3)$. If γ and ρ are primary, show that $-\gamma\rho$ is primary. If γ is primary, show that $\gamma = \pm\gamma_1\gamma_2\cdots\gamma_t$, where the γ_i are (not necessarily distinct) primary primes.

18 (continuation) If $\gamma = \pm\gamma_1\gamma_2\cdots\gamma_t$ is a primary decomposition of the primary element γ, define $\chi_\gamma(\alpha) = \chi_{\gamma_1}(\alpha)\chi_{\gamma_2}(\alpha)\cdots\chi_{\gamma_t}(\alpha)$. Prove that $\chi_\gamma(\alpha) = \chi_\gamma(\beta)$ if $\alpha \equiv \beta\ (\gamma)$ and $\chi_\gamma(\alpha\beta) = \chi_\gamma(\alpha)\chi_\gamma(\beta)$. If ρ is primary, show that $\chi_\rho(\alpha)\chi_\gamma(\alpha) = \chi_{-\rho\gamma}(\alpha)$.

19 Suppose that $\gamma = A + B\omega$ is primary and that $A = 3M - 1$ and $B = 3N$. Prove that $\chi_\gamma(\omega) = \omega^{M+N}$ and that $\chi_\gamma(\lambda) = \omega^{2M}$.

20 If γ and ρ are primary, show that $\chi_\gamma(\rho) = \chi_\rho(\gamma)$.

21 If γ is primary, show that there are infinitely many primary primes π such that $x^3 \equiv \gamma\ (\pi)$ is not solvable. Show also that there are infinitely many primary primes π such that $x^3 \equiv \omega\ (\pi)$ is not solvable and the same for $x^3 \equiv \lambda\ (\pi)$. (*Hint:* Imitate the proof of Theorem 3 of Chapter 5.)

22 (continuation) Show in general that if $\gamma \in D$ and $x^3 \equiv \gamma\ (\pi)$ is solvable for all but finitely many primary primes π, then γ is a cube in D.

23 Suppose that $p \equiv 1\ (3)$. Use Exercise 5 to show that $x^3 \equiv 3\ (p)$ is solvable in $\mathbb{Z}$ iff p is of the form $4p = C^2 + 243B^2$.

chapter ten/EQUATIONS OVER FINITE FIELDS

In this chapter we shall introduce a new point of view. Diophantine problems over finite fields will be put into the context of elementary algebraic geometry. The notions of affine space, projective space, and points at infinity will be defined.

After these problems of language have been dealt with, we shall prove a very general theorem due to C. Chevalley, which states that a polynomial in several variables with no constant term over a finite field always has nontrivial zeros if the number of variables exceeds the degree.

Next, our interest turns to the problem of generalizing the results of Chapter 8 *to arbitrary finite fields. This turns out to be relatively easy. These more general results are of interest for their own sake and are crucial to the discussion of the zeta function, which we shall take up in Chapter* 11.

I *AFFINE SPACE, PROJECTIVE SPACE, AND POLYNOMIALS*

Let F be a field and $A^n(F)$ the set of n-tuples $(a_1, a_2, \ldots, a_n)$ with $a_i \in F$. $A^n(F)$ can be considered as a vector space by defining addition and scalar multiplication in the usual way. We shall be concerned principally with the underlying set, which will be called affine n-space over F. As usual the point $(0, 0, \ldots, 0)$ will be called the origin. If there is no chance of confusion we shall denote the point $(a_1, a_2, \ldots, a_n)$ by the single letter a.

Projective n-space over F, $P^n(F)$, is a somewhat more difficult concept. We first construct $A^{n+1}(F)$, denoting its points by $(a_0, a_1, \ldots, a_n)$. On the set $A^{n+1}(F) - \{(0, 0, \ldots, 0)\}$ [affine $(n+1)$-space from which the origin has been removed] we define an equivalence relation. $(a_0, a_1, \ldots, a_n)$ is said to be equivalent to $(b_0, b_1, \ldots, b_n)$ if there is a $\gamma \in F^*$ such that $a_0 = \gamma b_0, a_1 = \gamma b_1, \ldots, a_n = \gamma b_n$. This is easily seen to be an equivalence relation. The equivalence classes are called points of $P^n(F)$. If $a \in A^{n+1}(F)$ is distinct from the origin, then $\bar{a}$ will denote the equivalence class containing a. a will be called a representative of $\bar{a}$. Geometrically, the points of $P^n(F)$ are in one-to-one correspondence with the lines in $A^{n+1}(F)$ that pass through the origin.

If F is a finite field with q elements, then clearly $A^n(F)$ has q^n elements. $P^n(F)$ has $q^n + q^{n-1} + \cdots + q + 1$ elements. To see this, notice that $A^{n+1}(F) - \{(0, 0, \ldots, 0)\}$ has $q^{n+1} - 1$ elements. Since F^* has $q - 1$ elements each equivalence class has $q - 1$ elements. Thus $P^n(F)$ has $(q^{n+1} - 1)/(q - 1) = q^n + q^{n-1} + \cdots + q + 1$ elements.

In general $P^n(F)$ has more points than $A^n(F)$. This is made more precise as follows. If $\bar{x} \in P^n(F)$ and $x_0 \neq 0$, set $\phi(\bar{x}) = (x_1/x_0, x_2/x_0, \ldots, x_n/x_0) \in A^n(F)$. This map is easily seen to be independent of the representative x.

Lemma 1

Let $\bar{H}$ be the set of $\bar{x} \in P^n(F)$ such that $x_0 = 0$. Then ϕ maps $P^n(F) - \bar{H}$ to $A^n(F)$ and this map is one to one and onto. (If S and R are sets, then $S - T$ is the set of elements in S but not in T.)

PROOF

If $\phi(\bar{x}) = \phi(\bar{y})$, then $x_i/x_0 = y_i/y_0$ for $i = 0, 1, \ldots, n$. Let $\gamma = y_0/x_0$. Then $\gamma x_i = y_i$ for $i = 0, 1, \ldots, n$ and so $\bar{x} = \bar{y}$.

If $v = (v_1, v_2, \ldots, v_n) \in A^n(F)$, set $w = (1, v_1, v_2, \ldots, v_n)$. Then $\phi(\bar{w}) = v$.

The set $\bar{H}$ is called the hyperplane at infinity. It is easy to see that $\bar{H}$ has the structure of $P^{n-1}(F)$. Thus $P^n(F)$ is made up of two pieces, one a copy of $A^n(F)$, called the finite points, and the other a copy of $P^{n-1}(F)$, called the points at infinity.

Notice that $P^0(F)$ consists of just one point. Thus $P^1(F)$ has only one point at infinity. Similarly $P^2(F)$ has a (projective) line at infinity, etc.

Now that affine space and projective space have been defined we take up the subject of polynomials and see how they determine sets called hypersurfaces.

Let $F[x_1, x_2, \ldots, x_n]$ be the ring of polynomials in n variables over F. If $f \in F[x_1, \ldots, x_n]$, then

$$f(x) = \sum_{(i_1, i_2, \ldots, i_n)} a_{i_1 i_2 \cdots i_n} x_1^{i_1} x_2^{i_2} \cdots x_n^{i_n}$$

where the sum is over a finite set of n-tuples of nonnegative integers $(i_1, i_2, \ldots, i_n)$, where $a_{i_1 i_2 \cdots i_n} \neq 0$. A polynomial of the form $x_1^{i_1} x_2^{i_2} \cdots x_n^{i_n}$ is called a monomial. Its total degree is defined to be $i_1 + i_2 + \cdots + i_n$; its degree in the variable x_m is defined as i_m. The degree of $f(x)$ is the maximum of the total degrees of monomials that occur in $f(x)$ with nonzero coefficients. The degree in x_m is the maximum of the degrees in x_m of monomials that occur in $f(x)$ with nonzero coefficients. Call these two numbers $\deg f(x)$ and $\deg_m f(x)$. Then

(a) $\deg f(x)g(x) = \deg f(x) \deg g(x)$.

(b) $\deg_m f(x)g(x) = \deg_m f(x) \deg_m g(x)$.

If all the monomials that occur in $f(x)$ have degree l, then $f(x)$ is said to be homogeneous of degree l.

For example, if $f(x) = 1 + x_1x_2 + x_2x_3 + x_4^3$, then $\deg f(x) = 3$, $\deg_1 f(x) = \deg_2 f(x) = \deg_3 f(x) = 1$, and $\deg_4 f(x) = 3$. $f(x)$ is not homogeneous, but $h(x) = x_1^3 + x_2^3 + x_3^3 + x_1x_2x_3$ is homogeneous of degree 3.

A homogeneous polynomial is sometimes called a form. A form of degree 2 is called a quadratic form, and one of degree 3 is called a cubic form, etc.

Suppose that K is a field containing F. If $f(x) \in F[x_1, x_2, \ldots, x_n]$ and $a \in A^n(K)$, we can substitute a_i for x_i and compute $f(a)$.

This shows that $f(x)$ defines a function from $A^n(K)$ to K by sending a to $f(a)$. A point $a \in A^n(K)$ such that $f(a) = 0$ is called a zero of $f(x)$.

If K is a finite field with q elements, then $x^q - x$ defines the zero function on $A^1(K)$. Thus it may happen that a nonzero polynomial gives rise to the zero function. This cannot happen when K is infinite (see the Exercises).

Let $f(x)$ be a nonzero polynomial and define $H_f(K) = \{a \in A^n(K) \mid f(a) = 0\}$. $H_f(K)$ is called the hypersurface defined by f in $A^n(K)$. When K is a finite field, $H_f(K)$ is a finite set and it is natural to ask for the number of points in $H_f(K)$. In Chapter 8 we dealt with a number of special cases of this problem.

We now wish to define a projective hypersurface. Let $h(x) \in F[x_0, x_1, \ldots, x_n]$ be a nonzero homogeneous polynomial of degree d. As before, K is a field containing F. For $\gamma \in K^*$ we have $h(\gamma x) = \gamma^d h(x)$. It follows that if $a \in A^{n+1}(K)$ and $h(a) = 0$, then $h(\gamma a) = 0$. Thus we may define $\bar{H}_h(K) = \{\bar{a} \in P^n(K) \mid h(a) = 0\}$. This set is called the hypersurface defined by h in $P^n(K)$. Again, if K is finite, we can ask for the number of points in $\bar{H}_h(K)$.

It turns out that working with projective space leads to more unified results than working with affine space. We shall illustrate this point after defining the projective closure of an affine hypersurface.

Let $f(x) \in F[x_1, x_2, \ldots, x_n]$, and define $\bar{f}(y) = \bar{f}(y_0, y_1, \ldots, y_n)$ by

$$\bar{f}(y) = y_0^{\deg f} f\left(\frac{y_1}{y_0}, \frac{y_2}{y_0}, \ldots, \frac{y_n}{y_0}\right)$$

We shall see in a moment that $\bar{f}$ is a homogeneous polynomial. It will give rise to a hypersurface in $P^n(K)$. Roughly speaking, the new hypersurface will be obtained by $H_f(K)$ by adding points at infinity.

Lemma 2

$\bar{f}(y)$ is a homogeneous polynomial of degree equal to $\deg f$. Moreover, $\bar{f}(1, y_1, y_2, \ldots, y_n) = f(y_1, y_2, \ldots, y_n)$.

PROOF

Set $d = \deg f$ and consider a monomial $x_1^{i_1}x_2^{i_2}\cdots x_n^{i_n}$ of degree $l \le d$. Then $y_0^d(y_1/y_0)^{i_1}\cdots(y_n/y_0)^{i_n} = y_0^{d-l}y_1^{i_1}y_2^{i_2}\cdots y_n^{i_n}$, which is of degree d. Thus in $\bar{f}(y)$ all the monomials have degree d, which proves the first statement.

The second statement is immediate from the definition.

As examples, if $f(x) = x_1^3 + x_2^3 - 1$, then $\bar{f}(y) = y_1^3 + y_2^3 - y_0^3$; if $f(x) = 1 + 2x_1^3 - 3x_2^2$, then $\bar{f}(y) = y_0^3 + 2y_1^3 - 3y_0y_2^2$.

Consider the hypersurface $H_f(K) \subset A^n(K)$. $\bar{f}(y)$ is homogeneous in the variables $y_0, y_1, \ldots, y_n$ and so $\bar{f}$ defines a hypersurface $\bar{H}_{\bar{f}}(K)$ in $P^n(K)$. This projective hypersurface is called the projective closure of $H_f(K)$ in $P^n(K)$.

Let $\lambda : A^n(K) \to P^n(K)$ by $\lambda(a_1, a_2, \ldots, a_n) = \overline{(1, a_1, a_2, \ldots, a_n)}$. λ is one to one and moreover the image of $H_f(K)$ under λ is contained in $\bar{H}_{\bar{f}}(K)$ since clearly $\bar{f}(\overline{(1, a_1, \ldots, a_n)}) = f(a_1, a_2, \ldots, a_n) = 0$ for all $a \in H_f(K)$. In general $\bar{H}_{\bar{f}}(K)$ has more points than $H_f(K)$, namely, the intersection of $\bar{H}_{\bar{f}}(K)$ with the hyperplane at infinity.

All this will become clearer by means of examples, but before giving some we recall the definitions of the maps ϕ and λ and give a diagrammatic picture of $P^n(K)$:

$$\lambda : A^n(K) \to P^n(K) \quad \text{by} \quad \lambda(a_1, a_2, \ldots, a_n) = \overline{(1, a_1, a_2, \ldots, a_n)}$$

$$\phi : P^n(K) - \bar{H} \to A^n(K) \quad \text{by}$$

$$\phi(\overline{(b_0, b_1, \ldots, b_n)}) = \left(\frac{b_1}{b_0}, \frac{b_2}{b_0}, \ldots, \frac{b_n}{b_0}\right)$$

$P^n(K)$

$\operatorname{im} \lambda \approx A^n(K)$ Finite points	$\bar{H} \approx P^{n-1}(K)$ Points at infinity

Examples

1. $f(x) = x_1^2 + x_2^2 - 1$ over the field $F = \mathbb{Z}/p\mathbb{Z}$.

We have seen in Chapter 8, Section 3, that $f(x) = 0$ has $p - 1$ solutions if $p \equiv 1\ (4)$ and $p + 1$ solutions if $p \equiv 3\ (4)$.

$\bar{f}(y) = y_1^2 + y_2^2 - y_0^2$. The solutions $\overline{(p_0, p_1, p_2)}$, where $p_0 \neq 0$ corresponds to the affine solution $(p_1/p_0, p_2/p_0)$. Suppose that $\overline{(0, p_1, p_2)}$ is a solution. Then $p_1^2 + p_2^2 = 0$ or $(p_2/p_1)^2 = -1$. If $p \equiv 1\ (4)$, there is an $a \in F$ such that $a^2 = -1$ and in this case there are two points at infinity, namely, $\overline{(0, 1, a)}$ and $\overline{(0, 1, -a)}$. If $p \equiv 3\ (4)$, there is no $a \in F$

such that $a^2 = -1$ and consequently there are no points at infinity. In both cases, then, the hypersurface $\overline{H}_{\bar{f}}(F)$ has exactly $p + 1$ points.

2. $f(x) = x_1^n + x_2^n - 1$ over $F = \mathbb{Z}/p\mathbb{Z}$, where $p \equiv 1\ (n)$.

We have $\bar{f}(y) = y_1^n + y_2^n - y_0^n$. Thus the points at infinity on $\overline{H}_{\bar{f}}(K)$ are of the form $\overline{(0, y_1, y_2)}$, where $y_1^n + y_2^n = 0$. If -1 is not an nth power in F, then there are no points at infinity. If $a^n = -1$ for some $a \in F$, then there are n solutions to $x^n = -1$ in F [this follows from Proposition 4.2.1 since $p \equiv 1\ (n)$]. Call these solutions $a_1 = a, a_2, \ldots, a_n$. Then $\overline{(0, 1, a_1)}, \ldots, \overline{(0, 1, a_n)}$ are the points at infinity that are zeros of $\bar{f}(y)$. In the notation of Chapter 8, Section 4, the number of points at infinity is $\delta_n(-1)n$, and $N(x_1^n + x_2^n = 1) + \delta_n(-1)n$ is the number of points on the projective hypersurface (curve) defined by $y_1^n + y_2^n - y_0^n = 0$. Since the number of points in $P^1(F)$ is $p + 1$ the formula in Proposition 8.4.1 can be interpreted in the following way: The number of points on the projective curve $y_1^n + y_2^n - y_0^n = 0$ over $\mathbb{Z}/p\mathbb{Z}$ differs from the number of points on the projective line by an error term that does not exceed $(n - 1)(n - 2)\sqrt{p}$.

This result is a special case of the so-called Riemann hypothesis for finite fields, which states, roughly, that over a finite field with q elements, the number of points on a projective curve differs from the number of points on the projective line by an error term that does not exceed twice the genus (a number associated with the curve) times $\sqrt{q}$.

Special cases of this result were proved by various authors: Gauss, G. Herglotz, Hasse, and Davenport. The theorem was proved in full generality by Weil.

3. $f(x) = x_1^2 + x_2^2 + \cdots + x_m^2 - 1$ over $F = \mathbb{Z}/p\mathbb{Z}$, where m is even.

The number of finite points is given by $p^{m-1} - (-1)^{(m/2)((p-1)/2)} \cdot p^{(m/2)-1}$ (see Proposition 8.6.1). Since $\bar{f}(y) = y_1^2 + y_2^2 + \cdots + y_m^2 - y_0^2$ the number of points at infinity is equal to the number of solutions to $y_1^2 + y_2^2 + \cdots + y_m^2 = 0$ in $P^{m-1}(F)$. The number of affine solutions is given by $N = p^{m-1} + (-1)^{(m/2)((p-1)/2)}(p - 1)p^{(m/2)-1}$ (see Exercise 19 in Chapter 8) so the number of projective solutions is

$$\frac{N - 1}{p - 1} = p^{m-2} + p^{m-3} + \cdots + p + 1 + (-1)^{(m/2)((p-1)/2)}p^{(m/2)-1}$$

Adding the number of finite solutions to the solutions at infinity yields

$$p^{m-1} + p^{m-2} + \cdots + p + 1$$

This result is rather remarkable. It says that the number of points on the projective hypersurface given by $y_1^2 + y_2^2 + \cdots + y_m^2 - y_0^2 = 0$ is exactly equal to the number of points in $P^{m-1}(\mathbb{Z}/p\mathbb{Z})$.

There is a simpler way to achieve this result. Instead of considering the finite and infinite points separately one simply counts the number M of affine solutions to $y_1^2 + y_2^2 + \cdots + y_m^2 - y_0^2 = 0$ in $A^{m+1}(F)$ and then calculates $(M - 1)/(p - 1)$. Since $m + 1$ is odd, the number M is equal to p^m (see Exercise 19 in Chapter 8). Thus $(M - 1)/(p - 1) = p^{m-1} + p^{m-2} + \cdots + p + 1$.

2 CHEVALLEY'S THEOREM

In this section F will denote a finite field with q elements.

If q is a prime, i.e., $F = Z/qZ$, the equation $x_1^{q-1} + x_2^{q-1} + \cdots + x_{q-1}^{q-1} = 0$ has no solution except $(0, 0, \ldots, 0)$ because a^{q-1} is equal to 1 or zero depending on whether $a \neq 0$ or $a = 0$ for $a \in F$. Thus the values taken on by the above polynomial are $0, 1, 2, \ldots, q - 1$ and it is zero only if $x_1 = x_2 = \cdots = x_{q-1} = 0$. Notice that for this polynomial the number of variables is equal to the degree.

In 1935 E. Artin conjectured the following theorem, which was proved almost immediately by C. Chevalley [16].

Theorem 1

Let $f(x) \in F[x_1, x_2, \ldots, x_n]$ and suppose that

(a) $f(0, 0, \ldots, 0) = 0$.

(b) $n > d = \deg f$.

Then f has at least two zeros in $A^n(F)$.

Before giving the proof we shall deduce an immediate corollary.

Corollary

Let $h(y) \in F[y_0, y_1, \ldots, y_n]$ be a homogeneous polynomial of degree $d > 0$. If $n + 1 > d$, then $\bar{H}_h(F)$ is not empty.

PROOF

Since h is homogeneous $(0, 0, \ldots, 0)$ is a zero. By Theorem 1 h has another zero, $(a_0, a_1, \ldots, a_n)$. Clearly $(a_0, a_1, \ldots, a_n) \in \bar{H}_h(K)$.

We shall need the following elementary lemmas.

Lemma 1

Let $f(x_1, x_2, \ldots, x_n)$ be a polynomial that is of degree less than q in each of its variables. Then if f vanishes on all of $A^n(F)$, it is the zero polynomial.

PROOF

The proof is by induction on n. If $n = 1$, $f(x)$ is a polynomial in one variable of degree less than q with q distinct roots, namely, all the elements of F. Thus f is identically zero.

Suppose that we have proved the result for $n - 1$ and consider $f(x_1, x_2, \ldots, x_n)$. We can write

$$f(x_1, \ldots, x_n) = \sum_{i=0}^{q-1} g_i(x_1, \ldots, x_{n-1})x_n^i$$

Select $a_1, a_2, \ldots, a_{n-1} \in F$. Then $\sum_{i=1}^{q-1} g_i(a_1, a_2, \ldots, a_{n-1})x_n^i$ has q roots and so $g_i(a_1, a_2, \ldots, a_{n-1}) = 0$. By induction each polynomial g_i is identically zero and hence so is f.

Remember that $f(x) = x^q - x$ is a nonzero polynomial that vanishes on all of $A^1(F)$, so the hypothesis of the lemma is crucial.

If a polynomial is of degree less than q in each variable, it is said to be reduced. Two polynomials f, g are said to be equivalent if $f(a) = g(a)$ for all $a \in A^n(F)$. We write $f \sim g$.

Lemma 2

Each polynomial $f(x) \in F[x_1, \ldots, x_n]$ is equivalent to a reduced polynomial.

PROOF

Consider the case of one variable. Clearly $x^q \sim x$. If $m > 0$ is an integer, let l be the least positive integer such that $x^m \sim x^l$. We claim that $l < q$. If not, $l = qs + r$ with $0 \leq r < q$ and $s \neq 0$. Then $x^l = (x^q)^s x^r \sim x^{s+r}$. Since $s + r < l$ this contradicts the minimality of l.

In the case of n variables consider the monomial $x_1^{i_1}x_2^{i_2}\cdots x_n^{i_n}$. By what has been said, $x_1^{i_1}x_2^{i_2}\cdots x_n^{i_n} \sim x_1^{j_1}x_2^{j_2}\cdots x_n^{j_n}$, where $j_k < q$ for $k = 1, 2, \ldots, n$. Lemma 2 follows directly from this remark.

We are now in a position to prove Theorem 1. Suppose that $(0, 0, \ldots, 0)$ is the only zero of f. Then $1 - f^{q-1}$ has the value 1 at $(0, 0, \ldots, 0)$ and the value zero elsewhere. The same is true of the polynomial $(1 - x_1^{q-1})(1 - x_2^{q-1})\cdots(1 - x_n^{q-1})$. Thus

$$1 - f^{q-1} - (1 - x_1^{q-1})(1 - x_2^{q-2})\cdots(1 - x_n^{q-1})$$

vanishes on all of $A^n(F)$. Replace $1 - f^{q-1}$ by an equivalent reduced polynomial g. Then

$$g - (1 - x_1^{q-1})\cdots(1 - x_n^{q-1})$$

is of degree less than q in each of its variables and vanishes on all of $A^n(F)$. By Lemma 1 it vanishes identically. Thus $\deg g = n(q - 1)$.

On the other hand, $\deg g \leq \deg(1 - f^{q-1}) = d(q - 1)$. Recall that $d = \deg f$. This implies that $n \leq d$, which is contrary to the hypothesis. Consequently f must have more than one zero.

We shall give another proof due to Ax [3]. It is based on the following lemma.

Lemma 3

Let $i_1, i_2, \ldots, i_n$ be nonnegative integers. Then unless each i_j is divisible by $q - 1$ and nonzero we have

$$\sum_{a \in A^n(F)} a_1^{i_1} a_2^{i_2} \cdots a_n^{i_n} = 0$$

PROOF

Suppose first that $n = 1$. If $i = 0$, then $\sum_{a \in F} a^0 = q = 0$ in F. Suppose $i \neq 0$. F^* is cyclic. Let b be a generator. If $q - 1 \nmid i$, then

$$\sum_{a \in F} a^i = \sum_{k=0}^{q-2} b^{ki} = \frac{b^{(q-1)i} - 1}{b^i - 1} = 0$$

In general

$$\sum_{a \in A^n(F)} a_1^{i_1} a_2^{i_2} \cdots a_n^{i_n} = \left(\sum_{a_1 \in F} a_1^{i_1}\right)\left(\sum_{a_2 \in F} a_2^{i_2}\right) \cdots \left(\sum_{a_n \in F} a_n^{i_n}\right)$$

Lemma 3 is now clear.

It should be remarked that if $q - 1 \mid i_j$ and $i_j \neq 0$ for all j, then the value of the above sum is $(q - 1)^n$.

To return to Theorem 1, let N_f be the number of solutions of $f(x) = 0$ in $A^n(F)$. We shall show that $p \mid N_f$, where p is the characteristic of F. This refinement of Chevalley's theorem was first given by Warning [78].

As we have seen, $1 - f^{q-1}$ has the value 1 at a zero of f and the value zero otherwise. Thus

$$\overline{N}_f = \sum_{a \in A^n(f)} (1 - f(a)^{q-1})$$

where $\overline{N}_f$ is the residue class of N_f mod p considered as an element of F.

Let $x_1^{i_1} x_2^{i_2} \cdots x_n^{i_n}$ be a monomial occurring in $1 - f(x)^{q-1}$. Since this polynomial has degree $d(q - 1)$ we must have $i_j < q - 1$ for some j since otherwise the degree of the monomial would exceed $n(q - 1)$ and we have assumed that $d < n$. By Lemma 3 $\sum_{a \in A^n(F)} a_1^{i_1} a_2^{i_2} \cdots a_n^{i_n} = 0$. Since $1 - f(x)^{q-1}$ is a linear combination of monomials it follows that $\overline{N}_f = 0$, or $p \mid N_f$.

Warning was able to prove that $N_f \geq q^{n-d}$. In a somewhat different direction Ax showed that $q^b \mid N_f$, where b is the largest integer less than n/d. See [78] and [3] for details.

3 GAUSS AND JACOBI SUMS OVER FINITE FIELDS

Let $\zeta_p = e^{2\pi i/p}$ and $F_p = \mathbb{Z}/p\mathbb{Z}$. In Chapter 8 the function $\psi : F_p \to C$ given by $\psi(t) = \zeta_p^t$ played a crucial role. To carry over the principal results of Chapter 8 to an arbitrary finite field F, we need an analog of ψ for F. This is done by means of the trace.

Suppose that F has $q = p^n$ elements. For $\alpha \in F$ define $\operatorname{tr}(\alpha) = \alpha + \alpha^p + \alpha^{p^2} + \cdots + \alpha^{p^{n-1}}$. $\operatorname{tr}(\alpha)$ is called the trace of α.

Proposition 10.3.1

If $\alpha, \beta \in F$ and $a \in F_p$, then

(a) $\operatorname{tr}(\alpha) \in F_p$.

(b) $\operatorname{tr}(\alpha + \beta) = \operatorname{tr}(\alpha) + \operatorname{tr}(\beta)$.

(c) $\operatorname{tr}(a\alpha) = a \operatorname{tr}(\alpha)$.

(d) tr *maps F onto F_p*.

PROOF

(a) We have

$$(\alpha + \alpha^p + \cdots + \alpha^{p^{n-1}})^p = \alpha^p + \alpha^{p^2} + \cdots + \alpha^{p^{n-1}} + \alpha^{p^n}$$

Since $\alpha^{p^n} = \alpha^q = \alpha$ we see that $\operatorname{tr}(\alpha)^p = \operatorname{tr}(\alpha)$. This proves property (a) (see Proposition 7.1.1, Corollary 1).

(b)
$$\begin{aligned}\operatorname{tr}(\alpha + \beta) &= (\alpha + \beta) + (\alpha + \beta)^p + \cdots + (\alpha + \beta)^{p^{n-1}} \\ &= (\alpha + \beta) + (\alpha^p + \beta^p) + \cdots + (\alpha^{p^{n-1}} + \beta^{p^{n-1}}) \\ &= (\alpha + \alpha^p + \cdots + \alpha^{p^{n-1}}) + (\beta + \beta^p + \cdots + \beta^{p^{n-1}}) \\ &= \operatorname{tr}(\alpha) + \operatorname{tr}(\beta)\end{aligned}$$

(c)
$$\begin{aligned}\operatorname{tr}(a\alpha) &= a\alpha + a^p\alpha^p + \cdots + a^{p^{n-1}}\alpha^{p^{n-1}} \\ &= a(\alpha + \alpha^p + \cdots + \alpha^{p^{n-1}}) \\ &= a \operatorname{tr}(\alpha)\end{aligned}$$

We have used the fact that $a^p = a$ for $a \in F_p$.

(d) The polynomial $x + x^p + \cdots + x^{p^{n-1}}$ has at most p^{n-1} roots in F. Since F has p^n elements there is an $\alpha \in F$ such that $\operatorname{tr}(\alpha) = c \neq 0$. If $b \in F_p$, then using property (c) we see that $\operatorname{tr}((b/c)\alpha) = (b/c)\operatorname{tr}(\alpha) = b$. Thus the trace is onto.

We now define $\psi : F \to C$ by the formula $\psi(\alpha) = \zeta_p^{\operatorname{tr}(\alpha)}$. If $F = F_p$, this coincides with the previous definition.

Proposition 10.3.2

The function ψ has the following properties:

(a) $\psi(\alpha + \beta) = \psi(\alpha)\psi(\beta)$.

(b) *There is an* $\alpha \in F$ *such that* $\psi(\alpha) \neq 1$.

(c) $\sum_{\alpha \in F} \psi(\alpha) = 0$.

PROOF

(a) $\psi(\alpha + \beta) = \zeta_p^{\operatorname{tr}(\alpha+\beta)} = \zeta_p^{\operatorname{tr}(\alpha)+\operatorname{tr}(\beta)} = \zeta_p^{\operatorname{tr}(\alpha)}\zeta_p^{\operatorname{tr}(\beta)} = \psi(\alpha)\psi(\beta)$.

(b) tr is onto, so there is an $\alpha \in F$ such that $\operatorname{tr}(\alpha) = 1$. Then $\psi(\alpha) = \zeta_p \neq 1$.

(c) Let $S = \sum_{\alpha \in F} \psi(\alpha)$. Choose β such that $\psi(\beta) \neq 1$. Then $\psi(\beta)S = \sum_{\alpha \in F} \psi(\beta)\psi(\alpha) = \sum_{\alpha \in F} \psi(\beta + \alpha) = S$. It follows that $S = 0$.

Proposition 10.3.3

Let $\alpha, x, y \in F$. *Then*

$$\frac{1}{q} \sum_{\alpha \in F} \psi(\alpha(x - y)) = \delta(x, y)$$

where $\delta(x, y) = 1$ *if* $x = y$ *and zero otherwise.*

PROOF

If $x = y$, then $\sum_{\alpha \in F} \psi(\alpha(x - y)) = \sum_{\alpha \in F} \psi(0) = q$.

If $x \neq y$, then $x - y \neq 0$ and $\alpha(x - y)$ ranges over all of F as α ranges over all of F. Thus $\sum_{\alpha \in F} \psi(\alpha(x - y)) = \sum_{\beta \in F} \psi(\beta) = 0$ by property (c) of Proposition 10.3.2.

Proposition 10.3.3 generalizes the corollary to Lemma 1 of Chapter 6.

In Chapter 7 we proved that the multiplicative group of a finite field is cyclic. On the basis of this fact, one easily sees that all the definitions and propositions of Chapter 8, Section 1, can be applied to F as well as to F_p. It is only necessary to replace p by q whenever it occurs. Thus we may assume that the theory of multiplicative characters for F is known.

We are now in a position to define Gauss sums on F.

Definition

Let χ be a character of F and $\alpha \in F^*$. Let $g_\alpha(\chi) = \sum_{t \in F} \chi(t)\psi(\alpha t)$. $g_\alpha(\chi)$ is called a *Gauss sum on F* belonging to the character χ.

If we replace p by q, Propositions 8.2.1 and 8.2.2 can now be proved for the sums $g_\alpha(\chi)$. In the proof of Proposition 8.2.2 one needs Proposition 10.3.3.

In particular, we have $|g_\alpha(\chi)| = q^{1/2}$ and $g_\alpha(\chi)g_\alpha(\chi^{-1}) = \chi(-1)q$ for $\chi \neq \varepsilon$.

The general theory of Jacobi sums and the interrelation between Gauss sums and Jacobi sums that is developed in Chapter 8, Section 5, generalizes with no difficulty (just replace p by q everywhere), and all the results of Chapter 8, Section 7, also hold. The reader may wish to go back to these sections to assure himself that there are indeed no difficulties in generalizing the definitions and results.

As an exercise in working with these new tools, we present a theorem that is really a reformulation of some of our earlier work. This theorem will also be of use in Chapter 11.

Theorem 2

Suppose that F is a field with q elements and $q \equiv 1\ (m)$. The homogeneous equation $a_0y_0^m + a_1y_1^m + \cdots + a_ny_n^m = 0$, $a_0, a_1, \ldots, a_n \in F$, defines a hypersurface in $P^n(F)$. The number of points on this hypersurface is given by

$$q^{n-1} + q^{n-2} + \cdots + q + 1$$
$$+ \frac{1}{q-1} \sum_{\chi_0, \chi_1, \ldots, \chi_n} \chi_0(a_0^{-1}) \cdots \chi_n(a_n^{-1}) J_0(\chi_0, \chi_1, \ldots, \chi_n) \tag{1}$$

where $\chi_i^m = \varepsilon$, $\chi_i \neq \varepsilon$, and $\chi_0\chi_1 \cdots \chi_n = \varepsilon$.

Moreover, under these conditions

$$\frac{1}{q-1} J_0(\chi_0, \chi_1, \ldots, \chi_n) = \frac{1}{q} g(\chi_0)g(\chi_1) \cdots g(\chi_n) \tag{2}$$

PROOF

The number of points N on the hypersurface in $A^{n+1}(F)$ defined by $a_0y_0^m + a_1y_1^m + \cdots + a_ny_n^m = 0$ is given by

$$q^n + \sum_{\chi_0, \chi_1, \ldots, \chi_n} \chi_0(a_0^{-1})\chi_1(a_1^{-1}) \cdots \chi_n(a_n^{-1}) J_0(\chi_0, \chi_1, \ldots, \chi_n)$$

where the characters χ_i are subject to the conditions stated in Theorem 2. This follows from Theorem 5 of Chapter 8. The number we are looking for is $(N - 1)/(q - 1)$ and this yields Equation (1).

By Proposition 8.5.1, part (c), we have

$$J_0(\chi_0, \chi_1, \ldots, \chi_n) = \chi_0(-1)(q-1)J(\chi_1, \chi_2, \ldots, \chi_n) \tag{3}$$

By Theorem 3 of Chapter 8

$$J(\chi_1, \chi_2, \ldots, \chi_n) = \frac{g(\chi_1)g(\chi_2) \cdots g(\chi_n)}{g(\chi_1\chi_2 \cdots \chi_n)} \tag{4}$$

Multiply the numerator and denominator of the right-hand side by $g(\chi_0)$. Since $\chi_0\chi_1 \cdots \chi_n = \varepsilon$, we have $g(\chi_0)g(\chi_1\chi_2 \cdots \chi_n) = \chi_0(-1)q$. Combining this comment, Equations (3) and (4) yield Equation (2).

Notes

There is a pleasant account of geometry over finite fields in the book *Excursions into Mathematics* [7]. The authors discuss affine, projective, and even hyperbolic geometry. There is also a short but useful bibliography.

Artin's conjecture on polynomials over finite fields was made much earlier by Dickson (On the Representations of Numbers by Modular Forms, *Bull. Am. Math. Soc.*, **15**, 1909, 338–347). The first proof we gave is the original proof of Chevalley [16]. The second proof is due to J. Ax [3] and is found in M. Greenberg [37] and Samuel [68]. E. Warning's proof of a sharper result can be found in his original paper [78] and in Borevich and Shafarevich [9].

A. Meyer, in 1884, was able to prove that a quadratic form over the rationals in five or more variables always has a rational zero if it has a real zero. Hasse was able to prove that the same result, suitably generalized, holds over any algebraic number field. E. Artin was led by this and other considerations to conjecture that over a certain class of number fields a form of degree d in $n > d^2$ variable always has a nontrivial zero. He also made conjectures of this nature over other types of fields. For a discussion of this area of research, see the paper of S. Lang [53]. The most recent exposition of the subject is the book of Greenberg [37], who includes a counterexample to Artin's conjecture for p-adic fields, discovered in 1966 by G. Terjanian. Other counterexamples were provided shortly thereafter by S. Shanuel. There is much left to discover in this area, which is one of the most fascinating in modern number theory.

One last remark. If one looks at the case where the ground field is the field of rational functions over a finite field, then the Artin conjecture mentioned above has been proved by Carlitz [11]. More precisely, let F be a finite field and $K = F(x)$. Then every form of degree d in more than d^2 variables has a nontrivial zero in K. The proof makes ingenious use of the theorem of Chevalley proved in this chapter. It is a special case of a general result of S. Lang.

Exercises

1 If K is an infinite field and $f(x_1, x_2, \ldots, x_n)$ is a polynomial with coefficients in K, show that f is not identically zero on $A^n(K)$. (*Hint:* Imitate the proof of Lemma 1 in Section 2.)

2 In Section 1 it was asserted that H, the hyperplane at infinity in $P^n(F)$, has the structure of $P^{n-1}(F)$. Verify this by constructing a one-to-one, onto map from $P^{n-1}(F)$ to H.

3 Suppose that F has q elements. Use the decomposition of $P^n(F)$ into finite points and points at infinity to give another proof of the formula for the number of points in $P^n(F)$.

4 The hypersurface defined by a homogeneous polynomial of degree 1, $a_0x_0 + a_1x_1 + a_2x_2 + \cdots + a_nx_n$, is called a hyperplane. Show that any hyperplane in $P^n(F)$ has the same number of elements as $P^{n-1}(F)$.

5 Let $f(x_0, x_1, x_2)$ be a homogeneous polynomial of degree n in $F[x_0, x_1, x_2]$. Suppose that not every zero of $a_0x_0 + a_1x_1 + a_2x_2$ is a zero of f. Prove that there are at most n common zeros of f and $a_0x_0 + a_1x_1 + a_2x_2$ in $P^2(F)$. In more geometric language this says that a curve of degree n and a line have at most n points in common unless the line is contained in the curve.

6 Let F be a field with q elements. Let $M_n(F)$ be the set of $n \times n$ matrices with coefficients in F. Let $Sl_n(F)$ be the subset of those matrices with determinant equal to one. Show that $Sl_n(F)$ can be considered as a hypersurface in $A^{n^2}(F)$. Find a formula for the number of points on this hypersurface. [*Answer:* $(q-1)^{-1}(q^n - 1)(q^n - q)\cdots(q^n - q^{n-1})$.]

7 Let $f \in F[x_0, x_1, x_2, \ldots, x_n]$. One can define the partial derivatives $\partial f/\partial x_0, \partial f/\partial x_1, \ldots, \partial f/\partial x_n$ in a formal way. Suppose that f is homogeneous of degree m. Prove that $\sum_{i=0}^{n} x_i(\partial f/\partial x_i) = mf$. This result is due to Euler. (*Hint:* Do it first for the case that f is a monomial.)

8 (continuation) If f is homogeneous, a point $\bar{a}$ on the hypersurface defined by f is said to be singular if it is simultaneously a zero of all the partial derivatives of f. If the degree of f is prime to the characteristic, show that a common zero of all the partial derivatives of f is automatically a zero of f.

9 If m is prime to the characteristic of F, show that the hypersurface defined by $a_0x_0^m + a_1x_1^m + \cdots + a_nx_n^m$ has no singular points.

10 A point on an affine hypersurface is said to be singular if the corresponding point on the projective closure is singular. Show that this is equivalent to the following definition. Let $f \in F[x_1, x_2, \ldots, x_n]$, not necessarily homogeneous, and $a \in H_f(F)$. Then a is singular if it is a common zero of $\partial f/\partial x_i$ for $i = 1, 2, \ldots, n$.

11 Show that the origin is a singular point on the curve defined by $y^2 - x^3 = 0$.

12 Show that the affine curve defined by $x^2 + y^2 + x^2y^2 = 0$ has two points at infinity and that both are singular.

13 Suppose that the characteristic of F is not 2, and consider the curve defined by $ax^2 + bxy + cy^2 = 1$, where $a, b, c \in F$. If $b^2 - 4ac \notin F$, show that there are no points at infinity in $P^2(F)$. If $b^2 - 4ac \in F$, show that there are one or two points at infinity depending on whether $b^2 - 4ac$ is zero. If $b^2 - 4ac = 0$, show that the point at infinity is singular.

14 Consider the curve defined by $y^2 = x^3 + ax + b$. Show that it has no singular points (finite or infinite) if $4a^3 + 27b^2 \neq 0$.

15 Let $\mathbb{Q}$ be the field of rational numbers and p a prime. Show that the form $x_0^{n+1} + px_1^{n+1} + p^2x_2^{n+1} + \cdots + p^nx_n^{n+1}$ has no zeros in $P^n(\mathbb{Q})$. (*Hint:* If $\bar{a}$ is a zero, one can assume that the components of a are integers and that they are not all divisible by p.)

16 Show by explicit calculation that every cubic form in two variables over $Z/2Z$ has a nontrivial zero.

17 Show that for each $m > 0$ and finite field F_q there is a form of degree m in m variables with no nontrivial zero. [*Hint:* Let $\omega_1, \omega_2, \ldots, \omega_m$ be a basis for F_{q^m} over F_q and show that $f(x_1, x_2, \ldots, x_m) = \prod_{i=0}^{m-1} (\omega_1^{q^i} x_1 + \cdots + \omega_m^{q^i} x_m)$ has the required properties.]

18 Let $g_1, g_2, \ldots, g_m \in F_q[x_1, x_2, \ldots, x_n]$ be homogeneous polynomials of degree d and assume that $n > md$. Prove that there is a nontrivial common zero. [*Hint:* Let f be as in Exercise 17 and consider the polynomial $f(g_1(x_1, \ldots, x_n), \ldots, g_m(x_1, \ldots, x_n))$.]

19 Characterize those extensions F_{p^n} of F_p that are such that the trace is identically zero on F_p.

20 Show that if $\alpha \in F_q$ has trace zero, then $\alpha = \beta - \beta^p$ for some $\beta \in F_q$.

21 Let ψ be a map from F_q to $\mathbb{C}^*$ such that $\psi(\alpha + \beta) = \psi(\alpha)\psi(\beta)$ for all $\alpha, \beta \in F_q$. Show that there is a $\gamma \in F_q$ such that $\psi(x) = \zeta^{\operatorname{tr}(\gamma x)}$ for all $x \in F_q$, where $\zeta = e^{2\pi i/p}$.

22 If $g_\alpha(\chi)$ is a Gauss sum on F, defined in Section 3, show that

(a) $g_\alpha(\chi) = \overline{\chi(\alpha)} g(\chi)$.

(b) $g(\chi^{-1}) = g(\bar{\chi}) = \chi(-1)\overline{g(\chi)}$.

(c) $|g_\alpha(\chi)| = q^{1/2}$.

(d) $g(\chi)g(\chi^{-1}) = \chi(-1)q$.

23 Suppose that f is a function mapping F to $\mathbb{C}$. Define $\hat{f}(s) = (1/q) \sum_t f(t)\overline{\psi(st)}$ and prove that $f(t) = \sum_s \hat{f}(s)\psi(st)$. The last sum is called the finite Fourier series expansion of f.

24 In Exercise 23 take f to be a nontrivial character χ and show that $\hat{\chi}(s) = (1/q)g_{-s}(\chi)$.

chapter eleven/THE ZETA FUNCTION

The zeta function of an algebraic variety has played a major role in recent developments in diophantine geometry.

In 1924 E. Artin introduced the notion of a zeta function for a certain class of curves defined over a finite field. By analogy with the classical Riemann zeta function he conjectured that the Riemann hypothesis was valid for the functions he had defined. In special cases he was able to prove this. Remarkably, results of this nature can already be found in the work of Gauss (naturally, Gauss stated his results differently from Artin). Weil was able to prove (in 1948) that the Riemann hypothesis for nonsingular curves over a finite field was true in general.

In 1949 Weil published a paper in the Bulletin of the American Mathematical Society *entitled "Numbers of Solutions of Equations over Finite Fields." In this paper he defined the zeta function of an algebraic variety and announced a number of conjectures. Weil had already proved the validity of his conjectures for curves. Here he establishes the same results for a class of projective hypersurfaces. We shall give an exposition of part of this material. Most of the necessary tools have already been developed. The main new result that is needed is the Hasse–Davenport relation between Gauss sums. Weil gave a proof of this relation that is substantially simpler than the original. We shall give a proof due to P. Monsky that is even simpler than Weil's, although it is far from trivial.*

The Weil conjectures provided a profound stimulus to the development of algebraic geometry. New techniques are now available that have provided the means to prove all but one of the conjectures. The Riemann hypothesis for algebraic varieties is still an open question.

We conclude the book with a proof of a conjecture stated by Gauss in the last entry in his mathematical diary.

I THE ZETA FUNCTION OF A PROJECTIVE HYPERSURFACE

In Chapter 7, Section 2, we showed that if $F = \mathbb{Z}/p\mathbb{Z}$ and $s \geq 1$ an integer, then there exists a field K containing F with p^s elements. The same result holds true in general. Namely, if F is a finite field with q elements and $s \geq 1$ an integer, then there exists a field F_s containing F with q^s elements (this is F_{q^s} in our former terminology). The proof of the general case is almost identical with that of the special case (see the Exercises to Chapter 7).

Now, let $f(y) \in F[y_0, y_1, \ldots, y_n]$ be a homogeneous polynomial and let N_s be the number of points on the projective hypersurface $\bar{H}_f(F_s) \subset P^n(F_s)$. In less fancy language, N_s is the number of zeros of f in $P^n(F_s)$. We wish to investigate the way in which the numbers N_s depend on s.

At the end of this section we shall prove that the number N_s depends only on s and not on the field F_s. This will follow once we show that any two fields containing F and of the same dimension over F are isomorphic.

To study the numbers N_s we introduce the power series $\sum_{s=1}^{\infty} N_s u^s$. In all that follows it is possible to deal only with formal power series and thus to avoid all questions of convergence. To those who are uncomfortable with that notion, notice that $N_s \leq q^{s+1} - 1/q - 1 < q^{s+1}$. It follows that our series converges for all complex numbers u such that $|u| < q^{-1}$ and defines an analytic function in that disc.

Let $\exp u = \sum_{s=0}^{\infty} (1/s!)u^s$.

Definition

The *zeta function* of the hypersurface defined by f is the series given by

$$Z_f(u) = \exp\left(\sum_{s=1}^{\infty} \frac{N_s u^s}{s}\right)$$

It is possible to regard $Z_f(u)$ either as a formal power series or as a function of a complex variable defined and analytic on the disc $\{u \in \mathbb{C} \mid |u| < q^{-1}\}$.

It may seem strange to deal with $Z_f(u)$ instead of directly considering the series $\sum_{s=1}^{\infty} N_s u^s$. The reasons are mainly historical, although as we shall see the zeta function is, in fact, easier to handle.

As a first example, consider the hyperplane at infinity. By definition this is the set of points $(a_0, a_1, \ldots, a_n) \in P^n(F)$ with $a_0 = 0$. It is defined by the equation $x_0 = 0$. As we pointed out in Chapter 10 it is easy to see that $\bar{H}_{x_0}(F_s)$ has the same number of points as $P^{n-1}(F_s)$; that is,

$$N_s = q^{s(n-1)} + q^{s(n-2)} + \cdots + q^s + 1$$

It follows that

$$\sum_{s=1}^{\infty} \frac{N_s u^s}{s} = \sum_{m=0}^{n-1} \left(\sum_{s=1}^{\infty} \frac{(q^m u)^s}{s} \right) = - \sum_{m=0}^{n-1} \ln (1 - q^m u) \tag{1}$$

We have used the identity $\sum_{s=1}^{\infty} w^s/s = -\ln(1 - w)$. Exponentiating Equation (1) yields

$$Z_{x_0}(u) = (1 - q^{n-1}u)^{-1}(1 - q^{n-2}u)^{-1} \cdots (1 - qu)^{-1}(1 - u)^{-1}$$

In particular, we see that $Z_{x_0}(u)$ is a rational function of u.

We shall now compute a somewhat more involved example. Consider the hypersurface defined by $-y_0^2 + y_1^2 + y_2^2 + y_3^2 = 0$. To compute N_1 we use Theorem 2 of Chapter 10. Specializing to our case we find that

$$N_1 = q^2 + q + 1 + \chi(-1)\frac{1}{q}g(\chi)^4$$

where χ is the character of order 2 on F. We know that $g(\chi)^2 = \chi(-1)q$. Thus

$$N_1 = q^2 + q + 1 + \chi(-1)q$$

To compute N_s we must replace q by q^s and χ by χ_s, the character of order 2 on F_s. Then

$$N_s = q^{2s} + q^s + 1 + \chi_s(-1)q^s$$

If -1 is a square in F, then $\chi_s(-1) = 1$ for all s. If -1 is not a square in F_s, it is not hard to see that $\chi_s(-1) = -1$ for s odd and $\chi_s(-1) = 1$ for s even.

In the first case

$$\sum_{s=1}^{\infty} \frac{N_s u^s}{s} = \sum_{s=1}^{\infty} \frac{(q^2u)^s}{s} + 2\sum_{s=1}^{\infty} \frac{(qu)^s}{s} + \sum_{s=1}^{\infty} \frac{u^s}{s}$$

and so

$$Z(u) = (1 - q^2u)^{-1}(1 - qu)^{-2}(1 - u)^{-1}$$

In the second case the last term gives rise to the sum

$$\sum_{s=1}^{\infty} \frac{(-qu)^s}{s} = -\ln(1 + qu)$$

Thus in this case

$$Z(u) = (1 - q^2u)^{-1}(1 - qu)^{-1}(1 + qu)^{-1}(1 - u)^{-1}$$

Notice that in the first case the zeta function has a pole at $u = q^{-1}$ of order 2, whereas in the second case there is a pole at $u = q^{-1}$ of order 1. This is in accordance with a conjecture of John Tate, which relates the order of the pole at $u = q^{-1}$ to certain geometric properties of the hypersurface. We cannot go more deeply into this here.

As a final example, consider the curve $y_0^3 + y_1^3 + y_2^3 = 0$ over $F = \mathbb{Z}/p\mathbb{Z}$, p a prime congruent to 1 modulo 3.

Specializing Theorem 2 of Chapter 10 once again we find that

$$N_1 = p + 1 + \frac{1}{p}g(\chi)^3 + \frac{1}{p}g(\chi^2)^3$$

Here χ is a cubic character on $\mathbb{Z}/p\mathbb{Z}$. We know that $g(\chi)^3 = p\pi$, where $\pi = J(\chi, \chi)$, and $\pi\bar{\pi} = p$. Thus

$$N_1 = p + 1 + \pi + \bar{\pi}$$

It will follow from the Hasse–Davenport relation, to be proved later, that

$$N_s = p^s + 1 - (-\pi)^s - (-\bar{\pi})^s$$

Calculation now shows that

$$\mathbb{Z}_f(u) = \frac{(1 + \pi u)(1 + \bar{\pi}u)}{(1 - u)(1 - pu)}$$

The numerator can be evaluated explicitly. In Chapter 8 we proved that $\pi + \bar{\pi} = A$, where A is uniquely determined by $4p = A^2 + 27B^2$ and $A \equiv 1$ (3).

So our final expression is

$$Z_f(u) = \frac{1 + Au + pu^2}{(1 - u)(1 - pu)}$$

In this example $Z_f(u)$ is a rational function; the numerator and denominator are polynomials with integer coefficients. The roots of $Z_f(u)$, π^{-1} and $\bar{\pi}^{-1}$, both have absolute value $p^{-1/2}$.

More generally, let $f(x_0, x_1, x_2) \in F[x_0, x_1, x_2]$ be a nonzero homogeneous polynomial that is absolutely nonsingular. Then, Weil was able to prove that the zeta function of f has the form

$$\frac{P(u)}{(1 - u)(1 - qu)}$$

where $P(u)$ is a polynomial with integer coefficients of degree $(d - 1)(d - 2)$, d being the degree of f. Furthermore, if α is a root of $P(u)$, then $|\alpha| = q^{-1/2}$. The last statement is called the Riemann hypothesis for curves.

[To see the relation with the classical Riemann hypothesis, make the change of variables $u = q^{-s}$ and set $\zeta_f(s) = Z_f(q^{-s})$. $\zeta_f(s)$ is directly analogous to the classical zeta function. The condition that the roots of $Z_f(u)$ have absolute value $q^{-1/2}$ is equivalent to the condition that the roots of $\zeta_f(s)$ have real part $\frac{1}{2}$.]

In all our examples the zeta function is rational. In 1959 B. Dwork proved that any algebraic variety has a rational zeta function [26]. His proof is extremely beautiful, but unfortunately it is based on methods that are beyond the scope of this book.

Our examples suggest another characterization of the condition that the zeta function is rational.

It is immediate from the definition of the zeta function that if it is expanded in a power series about the origin, then the constant term is 1. Consequently, if $Z_f(u) = P(u)/Q(u)$, where $P(u)$ and $Q(u)$ are polynomials, we may assume that $P(0) = Q(0) = 1$ (prove it). With this assumption, the zeta function can be factored as follows:

$$Z_f(u) = \frac{\prod_i (1 - \alpha_i u)}{\prod_j (1 - \beta_j u)}$$

where $\alpha_i, \beta_j \in \mathbb{C}$. We can now prove

Proposition 11.1.1

The zeta function is rational iff there exist complex numbers α_i and β_j such that

$$N_s = \sum_j \beta_j^s - \sum_i \alpha_i^s$$

PROOF

Suppose that the zeta function is rational. Then by the above remarks

$$Z(u) = \frac{\prod_i (1 - \alpha_i u)}{\prod_j (1 - \beta_j u)}$$

with $\alpha_i, \beta_j \in \mathbb{C}$. Taking logarithms

$$\ln Z(u) = \sum_i \ln(1 - \alpha_i u) - \sum_j \ln(1 - \beta_j u)$$

Differentiate both sides:

$$\frac{Z'(u)}{Z(u)} = \sum_i \frac{-\alpha_i}{1 - \alpha_i u} - \sum_j \frac{-\beta_j}{1 - \beta_j u}$$

Multiply both sides by u and then use the geometric series to expand in a power series. One finds finally that

$$\frac{uZ'(u)}{Z(u)} = \sum_{s=1}^{\infty} \left(\sum_j \beta_j^s - \sum_i \alpha_i^s \right) u^s \tag{2}$$

We now compute the left-hand side in a different way. From the definition

$$\ln Z(u) = \sum_{s=1}^{\infty} \frac{N_s u^s}{s}$$

Differentiate both sides and then multiply both sides by u. We find that

$$\frac{uZ'(u)}{Z(u)} = \sum_{s=1}^{\infty} N_s u^s \tag{3}$$

Comparing coefficients of u^s in Equations (2) and (3) we have

$$N_s = \sum_j \beta_j^s - \sum_i \alpha_i^s$$

The converse is an easy calculation that we have done in special cases. We leave the details to the reader.

It remains to prove that the numbers N_s are independent of the choice of field F_s. The reader may wish to simply accept this fact and proceed to Section 2.

Suppose that E and E' are two fields containing F both with q^s elements.

Proposition 11.1.2

E and E' are isomorphic over F; i.e., there exists a map $\sigma: E \to E'$ such that

(a) *σ is one to one and onto.*
(b) *$\sigma(a) = a$ for all $a \in F$.*
(c) *$\sigma(\alpha + \beta) = \sigma(\alpha) + \sigma(\beta)$ for all $\alpha, \beta \in E$.*
(d) *$\sigma(\alpha\beta) = \sigma(\alpha)\sigma(\beta)$ for all $\alpha, \beta \in E$.*

PROOF

We shall show that both E and E' are isomorphic over F to $F[x]/(f(x))$ for some irreducible polynomial $f(x) \in F[x]$.

To begin with there is an $\alpha' \in E'$ such that $E' = F(\alpha')$ (for example, take α' to be a primitive $q^s - 1$ root of unity). Let $f(x) \in F[x]$ be the monic irreducible polynomial for α'. Then $E' \approx F[x]/(f(x))$. Since α' satisfies $x^{q^s} - x = 0$ we have $f(x) \mid x^{q^s} - x$.

Since E has q^s elements we have $x^{q^s} - x = \prod_{\alpha \in E} (x - \alpha)$. It follows that $f(\alpha) = 0$ for some $\alpha \in E$.

Thus $F(\alpha) \approx F[x]/(f(x))$ is a subfield of E with q^s elements. One concludes that $E = F(\alpha) \approx F[x]/(f(x)) \approx F(\alpha') = E'$.

We can now use the isomorphism σ to induce a map $\bar{\sigma}$ from $P^n(E)$ to $P^n(E')$. Namely,

$$\bar{\sigma}(\overline{\alpha_0, \alpha_1, \ldots, \alpha_n}) = (\overline{\sigma(\alpha_0), \sigma(\alpha_1), \ldots, \sigma(\alpha_n)})$$

$\bar{\sigma}$ is one to one and onto. Moreover, if $f(y_0, y_1, \ldots, y_n) \in F[y_0, y_1, \ldots, y_n]$ and we restrict $\bar{\sigma}$ to the projective hypersurface $\bar{H}_f(E)$, it maps onto the projective hypersurface $\bar{H}_f(E')$. This proves the independence of the numbers N_s from the choice of field F_s. We leave the details to the reader.

2 TRACE AND NORM IN FINITE FIELDS

In Chapter 10, Section 3, we introduced the notion of trace. Here we shall generalize that notion and also define the norm in finite fields.

Let F be a finite field with q elements and E a field containing F with q^s elements.

Definition
If $\alpha \in E$, the *trace* of α from E to F is given by

$$\operatorname{tr}_{E/F}(\alpha) = \alpha + \alpha^q + \cdots + \alpha^{q^{s-1}}$$

The norm of α from E to F is given by

$$N_{E/F}(\alpha) = \alpha \cdot \alpha^q \cdots \alpha^{q^{s-1}}$$

The following two propositions describe the basic properties of trace and norm.

Proposition 11.2.1
If $\alpha, \beta \in E$ and $a \in F$, then

(a) $\operatorname{tr}_{E/F}(\alpha) \in F$.
(b) $\operatorname{tr}_{E/F}(\alpha + \beta) = \operatorname{tr}_{E/F}(\alpha) + \operatorname{tr}_{E/F}(\beta)$.
(c) $\operatorname{tr}_{E/F}(a\alpha) = a \operatorname{tr}_{E/F}(\alpha)$.
(d) $\operatorname{tr}_{E/F}$ *maps E onto F.*

Proposition 11.2.2
If $\alpha, \beta \in E$ and $a \in F$, then

(a) $N_{E/F}(\alpha) \in F$.
(b) $N_{E/F}(\alpha\beta) = N_{E/F}(\alpha)N_{E/F}(\beta)$.
(c) $N_{E/F}(a\alpha) = a^s N_{E/F}(\alpha)$.
(d) $N_{E/F}$ *maps E^* onto F^*.*

PROOF

The proof of Proposition 11.2.1 is exactly analogous to that of Proposition 10.3.1 and will be omitted.

To prove Proposition 11.2.2 notice that

$$N_{E/F}(\alpha)^q = (\alpha \cdot \alpha^q \cdot \cdots \cdot \alpha^{q^{s-1}})^q = \alpha^q \cdot \alpha^{q^2} \cdot \cdots \cdot \alpha^{q^s} = N_{E/F}(\alpha)$$

Thus $N_{E/F}(\alpha) \in F$.

Now,

$$\begin{aligned} N_{E/F}(\alpha\beta) &= (\alpha\beta) \cdot (\alpha\beta)^q \cdot \cdots \cdot (\alpha\beta)^{q^{s-1}} \\ &= (\alpha \cdot \alpha^q \cdot \cdots \cdot \alpha^{q^{s-1}}) \cdot (\beta \cdot \beta^q \cdot \cdots \cdot \beta^{q^{s-1}}) \\ &= N_{E/F}(\alpha) N_{E/F}(\beta) \end{aligned}$$

This proves step (b).

To prove step (c) notice that for $a \in F$, $N_{E/F}(a) = a \cdot a^q \cdot \cdots \cdot a^{q^{s-1}} = a^s$ since $a^q = a$. Now apply the result of step (b).

Finally, consider the kernel of the homomorphism $N_{E/F}$, i.e., the set of all $\alpha \in E$ such that $N_{E/F}(\alpha) = 1$. α is in the kernel iff

$$1 = \alpha \cdot \alpha^q \cdot \cdots \cdot \alpha^{q^{s-1}} = \alpha^{1+q+\cdots+q^{s-1}} = \alpha^{(q^s-1)/(q-1)}$$

Since $(q^s - 1)/(q - 1) \mid q^s - 1$ we have by Proposition 7.1.2 that $x^{(q^s-1)/(q-1)} = 1$ has $(q^s - 1)/(q - 1)$ solutions in E. By elementary group theory it follows that the image, $N_{E/F}(E^*)$, has $q - 1$ elements, but this is exactly the number of elements in F^*. Thus $N_{E/F}$ is onto.

Given a tower of fields $F \subset E \subset K$ we have the relation $[K:F] = [K:E][E:F]$. This result is easy to prove in general. If all three fields are finite, we can prove it as follows. Let q be the number of elements in F. Then the number of elements in E and K are $q^{[E:F]}$ and $q^{[K:F]}$, respectively. Considering K as an extension of E we can express the number of elements in K as $(q^{[E:F]})^{[K:E]}$. Thus

$$q^{[K:F]} = q^{[E:F][K:E]}$$

and therefore $[K:F] = [E:F][K:E]$.

We can now prove another simple property of trace and norm that will be useful.

Proposition 11.2.3

Let $F \subset E \subset K$ be three finite fields and $\alpha \in K$. Then

(a) $\mathrm{tr}_{K/F}(\alpha) = \mathrm{tr}_{E/F}(\mathrm{tr}_{K/E}(\alpha))$.

(b) $N_{K/F}(\alpha) = N_{E/F}(N_{K/E}(\alpha))$.

PROOF

We shall prove only property (a). The proof of property (b) is similar.

Let $d = [E : F]$, $m = [K : E]$, and $n = [K : F]$. As we have pointed out above, $n = dm$.

The number of elements in E is $q_1 = q^d$. Thus

$$\mathrm{tr}_{K/E}(\alpha) = \alpha + \alpha^{q_1} + \cdots + \alpha^{q_1^{m-1}}$$

and

$$\begin{aligned}
\mathrm{tr}_{E/F}(\mathrm{tr}_{K/E}(\alpha)) &= \sum_{i=0}^{d-1} \mathrm{tr}_{K/E}(\alpha)^{q^i} \\
&= \sum_{i=0}^{d-1} \sum_{j=0}^{m-1} \alpha^{q_1^j q^i} \\
&= \sum_{i=0}^{d-1} \sum_{j=0}^{m-1} \alpha^{q^{dj+i}} \\
&= \sum_{k=0}^{n-1} \alpha^{q^k} \\
&= \mathrm{tr}_{K/F}(\alpha)
\end{aligned}$$

We have used the fact that as j varies from zero to $m - 1$ and i varies from zero to $d - 1$ the quantity $dj + 1$ varies from zero to $md - 1 = n - 1$.

Suppose now that $F \subset K$ are finite fields, $n = [K : F]$, and $\alpha \in K$. Let $E = F(\alpha)$ and $f(x) \in F[x]$ be the minimal polynomial for α over F. By the corollary to Proposition 7.2.2 we have $[E : F] = d$, where d is the degree of $f(x)$.

Proposition 11.2.4

Write $f(x) = x^d - c_1 x^{d-1} + \cdots + (-1)^d c_d$. *Then*

(a) $f(x) = (x - \alpha)(x - \alpha^q) \cdots (x - \alpha^{q^{d-1}})$.

(b) $\mathrm{tr}_{K/F}(\alpha) = (n/d)c_1$.

(c) $N_{K/F}(\alpha) = c_d^{n/d}$.

PROOF

Since the coefficients of f satisfy $x^q = x$ we have

$$0 = f(\alpha)^q = f(\alpha^q)$$

Thus α^q is a root of f. Similarly,

$$0 = f(\alpha^q)^q = f(\alpha^{q^2})$$

Thus α^{q^2} is a root of f. Continuing in this manner we see that $\alpha, \alpha^q, \alpha^{q^2}, \ldots, \alpha^{q^{d-1}}$ are all roots of f. If we can show that all these roots are distinct, assertion (a) will follow.

Suppose that $0 \le i \le j < d$ and that $\alpha^{q^i} = \alpha^{q^j}$. Set $k = j - i$. We shall show that $k = 0$.

We have

$$\alpha^{q^i} = \alpha^{q^j} = (\alpha^{q^k})^{q^i}$$

which implies that

$$(\alpha - \alpha^{q^k})^{q^i} = 0$$

and so

$$\alpha = \alpha^{q^k}$$

Since $f(x)$ is the minimal polynomial for α it follows that $f(x)$ divides $x^{q^k} - x$ and so by Theorem 1 of Chapter 7 we have $d \mid k$. However, $0 \le k < d$ and so $k = 0$ and we are done.

It follows immediately from assertion (a) that $c_1 = \mathrm{tr}_{E/F}(\alpha)$ and that $c_d = N_{E/F}(\alpha)$.

Since $\alpha \in E = F(\alpha)$ we have $\mathrm{tr}_{K/E}(\alpha) = [K : E]\alpha = (n/d)\alpha$ and $N_{K/E}(\alpha) = \alpha^{n/d}$.

By Proposition 11.2.3,

$$\mathrm{tr}_{K/F}(\alpha) = \mathrm{tr}_{E/F}(\mathrm{tr}_{K/E}(\alpha)) = \mathrm{tr}_{E/F}\left(\frac{n}{d}\alpha\right) = \frac{n}{d}\mathrm{tr}_{E/F}(\alpha) = \frac{n}{d}c_1$$

Similarly,

$$N_{K/F}(\alpha) = N_{E/F}(N_{K/E}(\alpha)) = N_{E/F}(\alpha^{n/d}) = N_{E/F}(\alpha)^{n/d} = c_d^{n/d}$$

3 *THE RATIONALITY OF THE ZETA FUNCTION ASSOCIATED TO* $a_0x_0^m + a_1x_1^m + \cdots + a_nx_n^m$

Let $f(x_0, x_1, \ldots, x_n)$ be the polynomial given in the title of this section [notice that this is *not* the $f(x)$ of Section 2]. Suppose that the coefficients are in F, a finite field, with q elements and that $q \equiv 1\ (m)$. We have to investigate the number N_s of elements in $\bar{H}_f(F_s)$, where $[F_s : F] = s$. Theorem 2 of Chapter 10 shows that N_s is given by

$$q^{s(n-1)} + q^{s(n-2)} + \cdots + q^s + 1$$

$$+ \frac{1}{q^s} \sum_{\chi_0^{(s)}, \ldots, \chi_n^{(s)}} \chi_0^{(s)}(a_0^{-1}) \cdots \chi_n^{(s)}(a_n^{-1}) g(\chi_0^{(s)}) \cdots g(\chi_n^{(s)}) \qquad (4)$$

where q^s is the number of elements in F_s, and the $\chi_i^{(s)}$ are multiplicative characters of F_s such that $\chi_i^{(s)m} = \varepsilon$, $\chi_i^{(s)} \neq \varepsilon$, and $\chi_0^{(s)}\chi_i^{(s)} \cdots \chi_n^{(s)} = \varepsilon$.

We must analyze the terms $\chi_i^{(s)}(a_i^{-1})$ and $g(\chi_i^{(s)})$. To do this we first relate characters of F_s to characters of F.

Let χ be a character of F and set $\chi' = \chi \circ N_{F_s/F}$; i.e., for $\alpha \in F_s$, $\chi'(\alpha) = \chi(N_{F_s/F}(\alpha))$. Then one sees, using Proposition 11.2.2, that χ' is a character of F_s, and moreover that

(a) $\chi \neq \rho$ implies that $\chi' \neq \rho'$.
(b) $\chi^m = \varepsilon$ implies that $\chi'^m = \varepsilon$.
(c) $\chi'(a) = \chi(a)^s$ for all $a \in F$.

It follows easily that as χ varies over the characters of F of order dividing m, χ' varies over the characters of F_s of order dividing m.

The sum in Equation (4) can now be rewritten as

$$\sum_{\chi_0, \ldots, \chi_n} \chi_0(a_0^{-1})^s \cdots \chi_n(a_n^{-1})^s g(\chi_0') \cdots g(\chi_n') \tag{5}$$

where $\chi_0, \ldots, \chi_n$ are characters of F satisfying $\chi_i^m = \varepsilon$, $\chi_i \neq \varepsilon$, and $\chi_0\chi_1 \cdots \chi_n = \varepsilon$.

It remains to analyze the Gauss sums $g(\chi')$. This is the content of the following theorem of Hasse and Davenport (see [23]).

Theorem 1
$(-g(\chi))^s = -g(\chi')$.

We postpone the proof of this relation. Using Theorem 1 and Equations (4) and (5) we see that N_s is given by

$$\sum_{k=0}^{n-1} q^{ks} + (-1)^{n+1} \sum_{\chi_0, \chi_1, \ldots, \chi_n} \left[\frac{(-1)^{n+1}}{q} \chi_0(\alpha_0^{-1}) \cdots \chi_n(\alpha_n^{-1}) g(\chi_0) \cdots g(\chi_n)\right]^s \tag{6}$$

where the second sum is restricted by the same conditions as Equation (5).

Applying Proposition 11.1.1 gives us the main result of this chapter.

Theorem 2
Let $a_0, a_1, \ldots, a_n \in F$ be a finite field with q elements. Let $f(x_0, \ldots, x_n) = a_0x_0^m + a_1x_1^m + \cdots + a_nx_n^m$. Then the zeta function $Z_f(u)$ is a rational function of the form

$$\frac{P(u)^{(-1)^n}}{(1-u)(1-qu)\cdots(1-q^{n-1}u)}$$

where $P(u)$ is the polynomial

$$\prod_{\chi_0, \chi_1, \ldots, \chi_n} \left(1 - (-1)^{n+1} \frac{1}{q} \chi_0(a_0^{-1}) \cdots \chi_n(a_n^{-1}) g(\chi_0) g(\chi_1) \cdots g(\chi_n) u\right)$$

the $(n+1)$-tuples $\chi_0, \chi_1, \ldots, \chi_n$ being subject to the conditions $\chi_i^m = \varepsilon$, $\chi_i \neq \varepsilon$, and $\chi_0 \chi_1 \cdots \chi_n = \varepsilon$.

A number of remarks are in order:

1. The degree of $P(u)$ can be computed explicitly. It is

$$m^{-1}[(m-1)^{n+1} + (-1)^{n+1}(m-1)]$$

2. Since $|g(\chi)| = q^{1/2}$ it follows from the explicit expression for $P(u)$ that the zeros of $Z_f(u)$ have absolute value $q^{-((n-1)/2)}$. This is in accord with the general Riemann hypothesis.

3. If we write $P(u) = \prod (1 - \alpha u)$, then numbers α are algebraic integers. This is not hard to see. Each α has the form

$$\zeta \frac{1}{q} g(\chi_0) \cdots g(\chi_n)$$

where ζ is a root of unity and $\chi_0 \chi_1 \cdots \chi_n = \varepsilon$. Using Corollary 1 to Theorem 3 of Chapter 8 we see that

$$\frac{1}{q} g(\chi_0) g(\chi_1) \cdots g(\chi_n) = \chi_n(-1) J(\chi_0, \chi_1, \ldots, \chi_{n-1})$$

The Jacobi sum is a sum of roots of unity and so is an algebraic integer. Thus $\alpha = \zeta \chi_n(-1) J(\chi_0, \chi_1, \ldots, \chi_{n-1})$ is an algebraic integer as well.

4 *A PROOF OF THE HASSE–DAVENPORT RELATION*

Let F be a finite field with q elements and F_s be a field containing F such that $[F_s : F] = s$. Let χ be a nontrivial multiplicative character of F and $\chi' = \chi \circ N_{F_s/F}$. χ' is a character of F_s. We wish to compare the Gauss sums $g(\chi)$ and $g(\chi')$.

Let us recall the definition of $g(\chi)$ (see Chapter 10, Section 3):

$$g(\chi) = \sum_{t \in F} \chi(t) \psi(t)$$

where $\psi(t)$ is equal to $\zeta_p^{\mathrm{tr}(t)}$. The trace function in this definition coincides with the function tr_{F/F_p} introduced in this chapter. Since we are considering more than one field, it is important to attach subscripts to tr. Now,

$$g(\chi') = \sum_{t \in F_s} \chi'(t) \psi'(t)$$

where $\psi'(t) = \zeta_p^{\mathrm{tr}_{F_s/F_p}(t)}$. Since $\mathrm{tr}_{F_s/F_p}(t) = \mathrm{tr}_{F/F_p}(\mathrm{tr}_{F_s/F}(t))$ it follows that $\psi' = \psi \circ \mathrm{tr}_{F_s/F}$.

For a monic polynomial $f(x) = x^n - c_1 x^{n-1} + \cdots + (-1)^n c_n$ in $F[x]$ define $\lambda(f) = \psi(c_1)\chi(c_n)$.

Lemma 1

$\lambda(fg) = \lambda(f)\lambda(g)$ for all monic $f, g \in F[x]$.

PROOF

If $g(x) = x^m - b_1 x^{m-1} + \cdots + (-1)^m b_m$, then $f(x)g(x) = x^{n+m} - (b_1 + c_1)x^{n+m-1} + \cdots + (-1)^{n+m} b_m c_n$. Thus $\lambda(fg) = \psi(b_1 + c_1) \cdot \chi(b_m c_n) = \psi(b_1)\psi(c_1)\chi(b_m)\chi(c_n) = \psi(b_1)\chi(b_m)\psi(c_1)\chi(c_n) = \lambda(g)\lambda(f)$.

Lemma 2

Let $\alpha \in F_s$ and $f(x)$ be the monic irreducible polynomial for α over F. Then

$$\lambda(f)^{s/d} = \chi'(\alpha)\psi'(\alpha), \qquad \text{where } d = \deg f$$

PROOF

This result follows easily from Proposition 11.2.4. Namely, if $f(x) = x^d - c_1 x^{d-1} + \cdots + (-1)^d c_d$, then

$$\mathrm{tr}_{F_s/F}(\alpha) = \frac{s}{d} c_1 \quad \text{and} \quad N_{F_s/F}(\alpha) = c_d^{s/d}$$

Now, $\lambda(f) = \psi(c_1)\chi(c_d)$, so

$$\lambda(f)^{s/d} = \psi(c_1)^{s/d}\chi(c_d)^{s/d} = \psi\left(\frac{s}{d} c_1\right)\chi(c_d^{s/d})$$

$$= \psi(\mathrm{tr}_{F_s/F}(\alpha))\chi(N_{F_s/F}(\alpha)) = \psi'(\alpha)\chi'(\alpha)$$

Lemma 3

$g(\chi') = \sum (\deg f)\lambda(f)^{s/\deg f}$, where the sum is over all monic irreducible polynomials of $F[x]$ with degree dividing s.

PROOF

According to Theorem 1 of Chapter 7—generalized to F as base field—$x^{q^s} - x$ is the product of all monic irreducible polynomials of degree dividing s. It follows that every such irreducible polynomial has all its roots in F_s and conversely that every element in F_s satisfies such a polynomial.

Let $f(x)$ be monic irreducible of degree $d \mid s$. Let $\alpha_1, \alpha_2, \ldots, \alpha_d \in F_s$ be its roots. Then by Lemma 2

$$\sum_{i=1}^{d} \chi'(\alpha_i)\psi'(\alpha_i) = d\lambda(f)^{s/d}$$

Summing over all polynomials of the required type yields the result.

We are now in a position to prove the Hasse–Davenport relation. The proof is based on the following identity:

$$\sum_f \lambda(f)t^{\deg f} = \prod_f (1 - \lambda(f)t^{\deg f})^{-1} \tag{7}$$

where the sum is over all monic polynomials and the product is over all monic irreducible polynomials in $F[x]$.

The identity is proved by expanding each term $(1 - \lambda(f)t^{\deg f})^{-1}$ in a geometric series and using the fact that every monic polynomial can be written as the product of monic irreducible polynomials in a unique way. The details are left as an exercise.

Now,

$$\sum_f \lambda(f)t^{\deg f} = \sum_{s=0}^{\infty} \left(\sum_{\deg f = s} \lambda(f) \right) t^s$$

We define $\lambda(1) = 1$, as this is necessary for Equation (7) to hold. For $s = 1$ we have

$$\sum_{\deg f = 1} \lambda(f) = \sum_{a \in F} \lambda(x - a) = \sum_{a \in F} \chi(a)\psi(a) = g(\chi)$$

For $s > 1$ we have

$$\sum_{\deg f = s} \lambda(f) = \sum_{c_i \in F} \lambda(x^s - c_1 x^{s-1} + \cdots + (-1)^s c_s)$$

$$= q^{s-2} \sum_{c_1, c_s} \chi(c_1)\psi(c_s) = q^{s-2} \left(\sum_{c_1} \chi(c_1) \right) \left(\sum_{c_s} \psi(c_s) \right) = 0$$

Putting all this together we see that the left-hand side of Equation (7) reduces to $1 + g(\chi)t$. Using this, take the logarithm of both sides of Equation (7), differentiate, and multiply both sides of the result by t. This yields

$$\frac{g(\chi)t}{1 + g(\chi)t} = \sum_f \frac{\lambda(f)(\deg f)t^{\deg f}}{1 - \lambda(f)t^{\deg f}}$$

Expand the denominators in geometric series. Then

$$\sum_{s=1}^{\infty} (-1)^{s-1} g(\chi)^s t^s = \sum_f \left(\sum_{r=1}^{\infty} (\deg f)\lambda(f)^r t^{r \deg f} \right)$$

Equating the coefficients of t^s yields

$$(-1)^{s-1}g(\chi)^s = \sum_{\deg f|s} (\deg f)\lambda(f)^{s/\deg f}$$

By Lemma 3, the right-hand side is $g(\chi')$. This completes the proof.

5 THE LAST ENTRY

The last entry of Gauss's mathematical diary is a statement of the following remarkable conjecture:

Suppose that $p \equiv 1\ (4)$. Then the number of solutions to the congruence $x^2 + y^2 + x^2y^2 \equiv 1\ (p)$ is $p + 1 - 2a$, where $p = a^2 + b^2$ and $a + bi \equiv 1\ (2 + 2i)$.

Some explanation is in order. If $p \equiv 1\ (4)$, then by Proposition 8.3.1 we know that $p = a^2 + b^2$ for some integers a and b. If we choose a odd and b even, then a and b are uniquely determined up to sign. The congruence $a + bi \equiv 1\ (2 + 2i)$ determines the sign of a. We shall give a simpler formulation of this.

Lemma

If $p \equiv 1\ (4)$, $p = a^2 + b^2$, and $a + bi \equiv 1\ (2 + 2i)$, then a is odd and b is even. Moreover, if $4 \mid b$, then $a \equiv 1\ (4)$, and if $4 \nmid b$, then $a \equiv -1\ (4)$.

PROOF

$a + bi \equiv 1\ (2 + 2i)$ implies that $a + bi \equiv 1\ (2)$ and so a is odd and b even.

Since $4 = 2(i - 1)(i + 1)$ it follows that if $4 \mid b$, then $a + bi \equiv a \equiv 1\ (2 + 2i)$. Taking conjugates $a \equiv 1\ (2 - 2i)$. Thus $(2 + 2i)(2 - 2i) = 8 \mid (a - 1)^2$ and $a \equiv 1\ (4)$.

If $4 \nmid b$, then $b = 4k + 2$ for some k. Thus $a + bi \equiv a + 2i \equiv 1\ (2 + 2i)$. Since $2i \equiv -2\ (2 + 2i)$ we have $a \equiv 3 \equiv -1\ (2 + 2i)$. As before $8 \mid (a + 1)^2$ and so $a \equiv -1\ (4)$.

Theorem

Consider the curve C determined by $x^2t^2 + y^2t^2 + x^2y^2 - t^4$ over F_p, where $p \equiv 1\ (4)$. Write $p = a^2 + b^2$ with a odd and b even. If $4 \mid b$, choose $a \equiv 1\ (4)$; if $4 \nmid b$, choose $a \equiv -1\ (4)$. Then the number of points on C in $P^2(F_p)$ is $p - 1 - 2a$.

The zeta function of C is

$$Z(u) = \frac{1 - 2au + pu^2}{1 - pu}(1 - u)$$

Before giving the proof a few remarks are in order.

The answer $p - 1 - 2a$ differs from Gauss's $p + 1 - 2a$. The difficulty is that Gauss counts four points at infinity, whereas a simple calculation shows that $\overline{(0, 1, 0)}$ and $\overline{(0, 0, 1)}$ are the only points at infinity according to our definition. Thus our answer differs from his by 2.

Since there are two points at infinity independently of p it suffices to count the number of finite points, i.e., the solutions to $x^2 + y^2 + x^2y^2 = 1$.

As an example take $p = 5$. Since $5 = 1^2 + 2^2$ we have $4 \nmid b$ so we must take $a = -1$. The formula $p - 1 - 2a$ gives the answer 6 in this case. Indeed, in addition to the two points at infinity, $(1, 0)$, $(-1, 0)$, $(0, 1)$, and $(0, -1)$ are the other points on the curve in F_p.

The form of the zeta function may be surprising. The explanation is that the two points at infinity are singular. Thus the form of this zeta function is not in contradiction to our earlier observations.

We now proceed to prove the theorem. Denote by C_1 the curve given by $x^2 + y^2 + x^2y^2 = 1$ and by C_2 the curve given by $w^2 = 1 - z^4$. We shall construct maps from C_1 to C_2 and from C_2 to C_1.

Notice that

$$x^2 + y^2 + x^2y^2 = 1$$

implies that

$$(1 + x^2)y^2 = 1 - x^2$$

and

$$[(1 + x^2)y]^2 = 1 - x^4$$

Thus, if (a, b) is on C_1, then $(a, (1 + a^2)b)$ is on C_2. Let

$$\lambda(x, y) = (x, (1 + x^2)y)$$

λ maps C_1 to C_2. It is easy to see that this map is one to one.

Now let

$$\mu(z, w) = \left(z, \frac{w}{1 + z^2}\right)$$

μ is not always defined. If $\alpha \in F_p$ is such that $\alpha^2 = -1$, then $(\alpha, 0)$ and $(-\alpha, 0)$ are on C_2 but μ is undefined at these points. μ is defined at all other points of C_2 and maps these points to C_1. It is easy to check that μ is inverse to λ where it is defined. Thus

$$N_1 = N_2 - 2$$

where N_1 and N_2 are the number of finite points in F_p on C_1 and C_2, respectively.

We can compute N_2 by using Theorem 5 of Chapter 8. Specializing Theorem 5 to $w^2 + z^4 = 1$ we see that

$$N_2 = p + J(\rho, \chi) + J(\rho, \chi^2) + J(\rho, \chi^3)$$

where ρ is the character of order 2 and χ is a character of order 4.

Since $\chi^2 = \rho$, we have $J(\rho, \chi^2) = J(\rho, \rho) = -\rho(-1) = -1$. Also, since $\chi^4 = \varepsilon$ we have $\chi^3 = \bar{\chi}$ so that $J(\rho, \chi^3) = J(\rho, \bar{\chi}) = \overline{J(\rho, \chi)}$.

Let $\pi = -J(\rho, \chi)$. Then

$$N_2 = p - 1 - \pi - \bar{\pi}$$

ρ takes on the values ± 1 and χ takes on the values $\pm 1, \pm i$. Thus $\pi = a + bi$, where $a, b \in Z$. Moreover $|J(\rho, \chi)|^2 = p$ so that $a^2 + b^2 = \pi\bar{\pi} = p$. It follows that $N_2 = p - 1 - 2a$ and $N_1 = p - 3 - 2a$. Since C_1 has two points at infinity, the total number of points on C_1 in F_p is given by

$$N = p - 1 - 2a$$

By the lemma it suffices to prove that $\pi \equiv 1\ (2 + 2i)$ in order to complete the proof of the first part of the theorem. This is accomplished by means of the following pretty calculation given in Hasse–Davenport [23].

Notice that $\rho(a) - 1 \equiv 0\ (2)$ and that $\chi(a) - 1 \equiv 0\ (1 + i)$ for all $a \neq 0$ in F_p. The first assertion is obvious; the second follows from $1 - 1 = 0$, $-1 - 1 = -(1 - i)(1 + i)$, $-i - 1 = -(1 + i)$, and $i - 1 = i(1 + i)$. Thus if $a \neq 0$ and $b \neq 0$, $(\rho(a) - 1)(\chi(b) - 1) \equiv 0\,(2 + 2i)$. This congruence is trivially true for the pairs $a = 0$, $b = 1$ and $a = 1$, $b = 0$. Therefore,

$$\sum_{a+b=1} (\rho(a) - 1)(\chi(b) - 1) \equiv 0\,(2 + 2i)$$

Expanding we see that

$$-\pi - \sum_b \chi(b) - \sum_a \rho(a) + p \equiv 0\,(2 + 2i)$$

The second and third terms are zero. Thus

$$\pi \equiv p \equiv 1\,(2 + 2i)$$

The last step follows because $p \equiv 1\ (4)$ by hypothesis, and $2 + 2i$ divides 4; indeed $4 = (1 - i)(2 + 2i)$.

To calculate the zeta function it suffices to notice that by the Hasse–Davenport relation the number of points on $x^2t^2 + y^2t^2 + x^2y^2 - t^4$

in $P^2(F_{p^s})$ is given by

$$p^s - 1 - (-J(\rho, \chi))^s - (-J(\overline{\rho, \chi}))^s = p^s - 1 - \pi^s - \bar{\pi}^s$$

Thus

$$\begin{aligned} Z(u) &= \frac{(1 - \pi u)(1 - \bar{\pi} u)}{(1 - pu)}(1 - u) \\ &= \frac{1 - 2au + pu^2}{(1 - pu)}(1 - u) \end{aligned}$$

Notes

As we have mentioned, in his thesis E. Artin [2] introduced the congruence zeta function. In that work he establishes the analog of the Riemann hypothesis for about forty curves of the type $y^2 = f(x)$, where f is a cubic or quartic polynomial. In 1934 Hasse proved that the result held in general for nonsingular cubics and quartics (the case of elliptic curves). The Riemann hypothesis for arbitrary nonsingular curves was established in full generality by Weil in 1941. His proof is far from elementary and uses deep techniques in algebraic geometry. This result is one of the pinnacles of modern number theory.

Weil's conjecture that the zeta function of any algebraic set is rational was proved in 1959 by B. Dwork using methods of p-adic analysis [26]. The question about the absolute value of the roots of the zeta function associated with a nonsingular variety (generalized Riemann hypothesis) remains open, although Weil's conjectures in this direction have been verified in many particular cases. For an early survey paper in this area, see M. Deuring [24].

We have ignored the topological aspects of Weil's conjectures. Suffice it to say that they relate the form of the zeta function to certain topological invariants of the algebraic variety. This aspect of the conjectures led to the search for cohomology groups in characteristic p, but these matters are too technical to go into here.

Section 5 on Gauss's conjecture is logically out of place since it could have been given in Chapter 8. We felt it was appropriate at this point since the relation between this conjecture and Weil's Riemann hypothesis reveals once again the remarkable acuity of Gauss's insight and how his imposing presence continues to make itself felt to this very day.

Exercises

1 Suppose that we may write the power series $1 + a_1u + a_2u^2 + \cdots$ as the quotient of two polynomials $P(u)/Q(u)$. Show that we may assume that $P(0) = Q(0) = 1$.

2 Prove the converse to Proposition 11.1.1.

3 Give the details of the proof that N_s is independent of the field F_s (see the concluding paragraph to Section 1).

4 Calculate the zeta function of $x_0x_1 - x_2x_3 = 0$ over F_p.

5 Calculate as explicitly as possible the zeta function of $a_0x_0^2 + a_1x_1^2 + \cdots + a_nx_n^2$ over F_q, where q is odd. The answer will depend on whether n is odd or even and whether $q \equiv 1\ (4)$ or $q \equiv 3\ (4)$.

6 Consider $x_0^3 + x_1^3 + x_2^3 = 0$ as an equation over F_4, the field with four elements. Show that there are nine points on the curve in $P^2(F_4)$. Calculate the zeta function. [*Answer:* $(1 + 2u)^2/((1 - u)(1 - 4u))$.]

7 Try this exercise if you know a little projective geometry. Let N_s be the number of lines in $P^n(F_{p^s})$. Find N_s and calculate $\sum_{s=1}^{\infty} N_s u^s/s$. (The set of lines in projective space form an algebraic variety called a Grassmannian variety. So do the set of planes, three-dimensional linear subspaces, etc.)

8 If f is a nonhomogeneous polynomial, we can consider the zeta function of the projective closure of the hypersurface defined by f (see Chapter 10). One way to calculate this is to count the number of points on $H_f(F_q)$ and then add to it the number of points at infinity. For example, consider $y^2 = x^3$ over F_{p^s}. Show that there is one point at infinity. The origin $(0, 0)$ is clearly on this curve. If $x \neq 0$, write $(y/x)^2 = x$ and show that there are $p^s - 1$ more points on this curve. Altogether we have $p^s + 1$ points and the zeta function over F_p is $(1 - pu)^{-1}$.

9 Calculate the zeta function of $y^2 = x^3 + x^2$ over F_p.

10 If $A \neq 0$ in F_q and $q \equiv 1\ (3)$, show that the zeta function of $y^2 = x^3 + A$ over F_q has the form $Z(u) = (1 + au + qu^2)/(1 - u)(1 - qu)$, where $a \in Z$ and $|a| \leq 2q^{1/2}$.

11 Consider the curve $y^2 = x^3 - Dx$ over F_p, where $D \neq 0$. Call this curve C_1. Show that the substitution $x = \frac{1}{2}(u + v^2)$ and $y = \frac{1}{2}v(u + v^2)$ transforms C_1 into the curve C_2 given by $u^2 - v^4 = 4D$. Show that in any given finite field the number of finite points on C_1 is one more than the number of finite points on C_2.

12 (continuation) If $p \equiv 3\ (4)$, show that the number of projective points on C_1 is just $p + 1$. If $p \equiv 1\ (4)$, show that the answer is $p + 1 + \overline{\chi(D)J(\chi, \chi^2)} + \chi(D)J(\chi, \chi^2)$, where χ is a character of order 4 on F_p.

13 (continuation) If $p \equiv 1\ (4)$, calculate the zeta function of $y^2 = x^3 - Dx$ over F in terms of π and $\chi(D)$, where $\pi = -J(\chi, \chi^2)$. This calculation in somewhat sharpened form is contained in [23]. The result has played a key role in recent empirical work of B. J. Birch and H. P. F. Swinnerton-Dyer on elliptic curves.

14 Suppose that $p \equiv 1\ (4)$ and consider the curve $x^4 + y^4 = 1$ over F_p. Let χ be a character of order 4 and $\pi = -J(\chi, \chi^2)$. Give a formula for the number of projective points over F_p and calculate the zeta function. Both answers should depend only on π. (*Hint:* See Exercises 7 and 16 of Chapter 8, but be careful since there we were counting only finite points.)

15 Find the number of points on $x^2 + y^2 + x^2y^2 = 1$ for $p = 13$ and $p = 17$. Do it both by means of the formula in Section 5 and by direct calculation.

16 Let F be a field with q elements and F_s an extension of degree s. If χ is a character of F, let $\chi' = \chi \circ N_{F_s/F}$. Show that

(a) χ' is a character of F_s.

(b) $\chi \neq \rho$ implies that $\chi' = \rho'$.

(c) $\chi^m = \varepsilon$ implies that $\chi'^m = \varepsilon$.

(d) $\chi'(a) = \chi(a)^s$ for $a \in F$.

(e) As χ varies over all characters of F with order dividing m, χ' varies over all characters of F_s with order dividing m. Here we are assuming that $q \equiv 1\ (m)$.

17 In Theorem 2 show that the order of the numerator of the zeta function, $P(u)$, has degree $m^{-1}((m-1)^{n+1} + (-1)^{n+1}(m-1))$.

18 Let the notation be as in Exercise 16. Use the Hasse–Davenport relation to show that $J(\chi_1', \chi_2', \ldots, \chi_n') = (-1)^{(s-1)(n-1)} J(\chi_1, \chi_2, \ldots, \chi_n)^s$, where the χ_i are nontrivial characters of F and $\chi_1\chi_2\cdots\chi_n \neq \varepsilon$.

19 Prove the identity $\sum \lambda(f)t^{\deg f} = \prod (1 - \lambda(f)t^{\deg f})^{-1}$, where the sum is over all monic polynomials in $F[t]$ and the product is over all monic irreducibles in $F[t]$. λ is defined in Section 4.

20 If in Theorem 2 we consider the base field to be F_s instead of F, we get a different zeta function, $Z_f^{(s)}(u)$. Show that $Z_f^{(s)}(u)$ and $Z_f(u)$ are related by the equation $Z_f^{(s)}(u^s) = Z_f(u)Z_f'(\rho u)\cdots Z_f(\rho^{s-1}u)$, where $\rho = e^{2\pi i/s}$.

21 In Exercise 6 we considered the equation $x_0^3 + x_1^3 + x_2^3 = 0$ over the field with four elements. Consider the same equation over the field with two elements. The trouble here is that $2 \not\equiv 1\ (3)$ and so our usual calculations do not work. Prove that in every extension of $Z/2Z$ of odd degree every element is a cube and that in every extension of even degree, 3 divides the order of the multiplicative group. Use this information to calculate the zeta function over $Z/2Z$. [*Answer:* $(1 + 2u^2)/(1-u)(1-2u)$.]

22 Use the ideas developed in Exercise 21 to show that Theorem 2 continues to hold (in a suitable sense) even when the hypothesis $q \equiv 1\ (m)$ is removed.

BIBLIOGRAPHY

1 A. Albert, *Fundamental Concepts of Higher Algebra*. Chicago: University of Chicago Press, 1956.

2 E. Artin, *The Collected Papers of Emil Artin.* Reading, Mass.: Addison-Wesley Publishing Company, Inc., 1965.

3 J. Ax, Zeros of Polynomials over Finite Fields, *Am. J. Math.*, **86**, 225–261.

4 P. Bachman, *Niedere Zahlentheorie*, Vol. 1. Leipzig: 1902, p. 83.

5 P. Bachman, *Die Lehre von der Kreisteilung*. Leipzig: 1872.

6 P. Bachman, Über Gauss' Zahltheoretische Arbeiten, *Gott. Nach.*, 1911, 455–508.

7 A. Beck, M. N. Bleicher, and D. W. Grove, *Excursions into Mathematics*. New York: Worth Publishers, Inc., 1969.

8 H. Bilharz, Primdivisor mit vorgegebener Primitivwurzel, *Math. Ann.*, **114**, 1937, 476–492.

9 Z. I. Borevich and I. R. Shafarevich, *Number Theory*, trans. N. Greenleaf. New York: Academic Press, Inc., 1966.

10 L. Carlitz, The Arithmetic of Polynomials in a Galois Field, *Am. J. Math.*, **54**, 1932, 39–50.

11 L. Carlitz, Some Applications of a Theorem of Chevalley, *Duke J.*, **18**, 1951, 811–819.

12 L. Carlitz, Some Problems Involving Primitive Roots in a Finite Field, *Proc. Nat. Acad. Sci. U.S.A.*, **38**, 1952, 314–318.

13 L. Carlitz, Kloosterman Sums and Finite Field Extensions, *Acta Arithmetica*, **16**, 1969, 179–193.

14 P. Cartier, Sur une généralisation des symboles de Legendre–Jacobi, *L'Enseignement Math.*, **15**, 1970, 31–48.

15 J. W. S. Cassels, On Kummer Sums, *Proc. London Math. Soc.*, **21**, No. 3, 1970, 19–27.

16 C. Chevalley, Démonstration d'une hypothèse de M. Artin, *Abh. Math. Sém. der Univ. Hamburg*, **11**, 1936, 73–75.

17 S. Chowla, The Last Entry in Gauss' Diary, *Proc. Nat. Acad. Sci. U.S.A.*, **35**, 1949, 244–246.

18 S. Chowla, *The Riemann Hypothesis and Hilbert's Tenth Problem*. New York: Gorden & Breach, Science Publishers, Inc., 1965.

19 S. Chowla, A Note on the Construction of Finite Galois Fields $GF(p^n)$, *J. Math. Anal. Appl.*, **15**, 1966, 53–54.

20 S. Chowla, An Algebraic Proof of the Law of Quadratic Reciprocity, *Norske Vid. Selsk. Forh.* (*Trondheim*), **39**, 1966, 59.

21 H. Davenport, On the Distribution of Quadratic Residues mod p, *London Math. Soc. J.*, **5–6**, 1930–1931, 49–54.

22 H. Davenport, *The Higher Arithmetic*. London: Hutchinson Publishing Group Ltd., 1968.

23 H. Davenport and H. Hasse, Die Nullstellen der Kongruenz Zetafunktion in gewissen zyklischen Fallen, *J. Reine und Angew. Math.*, 1935, 151–182.

24 M. Deuring, The Zeta Functions of Algebraic Curves and Varieties, *Indian J. Math.*, 1955, 89–101.

25 L. Dickson, *Linear Algebraic Groups and an Exposition of the Galois Field Theory, 1900*. New York: Dover Publications, Inc., 1958.

26 B. Dwork, On the Rationality of the Zeta Function, *Am. J. Math.*, 1959, 631–648.

27 G. Eisenstein, Beitrage zum Kreisteilung, *J. Reine und Angew. Math.*, 1844, 269–278.

28 G. Eisenstein, Beweis der Reciprocitatssatzes für die kubische Reste . . . , *J. Reine und Angew. Math.*, 1844, 289–310.

29 G. Eisenstein, Nachtrag zum kubischen Reciprocitatssatze, *J. Reine und Angew. Math.*, **28**, 1844, 28–35.

30 G. Eisenstein, Beitrage zur Theorie der elliptischen Funktionen, *J. Reine und Angew. Math.*, **35**, 1847, 135–274.

31 P. Erdös, Some Recent Advances and Current Problems in Number Theory, *Lectures in Modern Mathematics*, Vol. 3. New York: John Wiley & Sons, Inc., 1965.

32 A. Frankel, Integers and the Theory of Numbers, *Scripta Math. Studies*, **5**, New York, 1955.

33 E. Galois, *Oeuvres mathématiques.* Paris: Gauthier-Villars, 1897.

34 C. F. Gauss, *Arithmetische Untersuchungen.* New York: Chelsea Publishing Company, Inc., 1965.

35 L. Goldstein, Density Questions in Algebraic Number Theory, *Am. Math. Monthly*, April 1971, 342–351.

36 R. Graham, On Quadruples of Consecutive k'th Power Residues, *Proc. Am. Math. Soc.*, 1964, 196–197.

37 M. Greenberg, *Forms in Many Variables.* Menlo Park, Calif.: W. A. Benjamin, Inc., 1969.

38 G. H. Hardy, *Prime Numbers*, British Association, Manchester, 1915, pp. 350–354.

39 G. H. Hardy, An Introduction to the Theory of Numbers, *Bull. Am. Math. Soc.*, **35**, 1929, 778–818.

40 G. H. Hardy and E. M. Wright, *An Introduction to the Theory of Numbers.* New York: Oxford University Press, Inc., 4th ed., 1960.

41 H. Hasse, *Vorlesungen über Zahlentheorie.* Berlin: Springer-Verlag, 1964.

42 H. Hasse, *The Riemann Hypothesis in Function Fields.* Philadelphia: University of Pennsylvania Press, 1969.

43 A. Hausner, On the Law of Quadratic Reciprocity, *Archiv der Math.*, **12**, 1961, 182–183.

44 E. Hecke, *Algebraische Zahlentheorie.* Leipzig: 1929, reprinted by Chelsea Publishing Company, Inc., New York.

45 L. Holzer, *Zahlentheorie.* Leipzig: Teubner Verlagsgesellschaft, 1958.

46 C. Hooley, On Artin's Conjecture, *J. Reine und Angew. Math.*, **225**, 1967, 209–220.

47 C. Jacobi, Uber die Kreisteilung . . . , *J. Reine und Angew. Math.*, 1846, 254–274.

48 E. Jacobsthal, Uber die Darstellungen der Primzahlen der Form $4n + 1$ als Summe zweier Quadrate, *J. Reine und Angew. Math.*, **132**, 1907, 238–245.

49 C. Jordan, *Traité des substitutions.* Paris: 1870.

50 H. Kornblum, Uber die Primfunktionen in einer Arithmetischen Progression, *Math. Z.*, **5**, 1919, 100–111.

51 E. Kummer, Über die allgemeinen Reciprocitatsgesetz . . . , *Math. Abh. Akad. Wiss. zu Berlin*, 1859, 19–160.

52 E. Landau, *Elementary Number Theory*. New York: Chelsea Publishing Company, Inc., 2nd ed., 1966.

53 S. Lang, Some Theorems and Conjectures on Diophantine Equations, *Bull. Am. Math. Soc.*, **66**, 1960, 240–249.

54 D. H. Lehmer, A Note on Primitives, *Scripta Mathematica*, **26**, 1963, 117–119.

55 E. Lehmer, On the Quintic Character of 2 and 3, *Duke Math. J.*, **18**, 1951, 11–18.

56 E. Lehmer, Criteria for Cubic and Quartic Residuacity, *Mathematika*, **6**, 1958, 20–29.

57 P. Leonard, On Constructing Quartic Extensions of $GF(p)$, *Norske Vid. Selsk. Forh. (Trondheim)*, **40**, 1967, 41–52.

58 H. B. Mann, *Introduction to Number Theory*. Columbus, Ohio: Ohio State University Press, 1955.

59 W. H. Mills, Bounded Consecutive Residues and Related Problems, *Proc. Symp. Pure Math.*, **8**, 1965.

60 T. Nagell, *Introduction to Number Theory*. New York: John Wiley & Sons, Inc., 1951; reprinted by Chelsea Publishing Company, Inc., New York.

61 I. Niven and H. S. Zuckerman, *An Introduction to the Theory of Numbers*. New York: John Wiley & Sons, Inc., 2nd ed., 1966.

62 C. Pisot, Introduction à la theorie des nombres algébriques, *L'Enseignement Math.*, **8**, No. 2, 1962, 238–251.

63 H. Pollard, *The Theory of Numbers*, New York: John Wiley & Sons, Inc., 1950.

64 H. Rademacher, *Lectures on Elementary Number Theory*. Lexington, Mass.: Xerox College Publishing, 1964.

65 H. Rademacher and O. Toeplitz, *The Enjoyment of Mathematics*. Princeton, N.J.: Princeton University Press, 1951.

66 G. Rieger, *Die Zahlentheorie bei C. F. Gauss*, from *Gauss Gedenkband*. Berlin: Haude and Sperner, 1960.

67 P. Samuel, Unique Factorization, *Am. Math. Monthly*, **75**, 1968, 945–952.

68 P. Samuel, *Théorie algébrique des nombres*. Paris: Hermann & Cie, 1967.

69 J. P. Serre, *Compléments d'arithmétiques, rédigés par J. P. Ramis et G. Ruget*, Ecoles Normales Supérieures, Paris, 1964.

70 D. Shanks, *Solved and Unsolved Problems in Number Theory*. New York: Spartan Books, 1962.

71 W. Sierpinski, *A Selection of Problems in the Theory of Numbers*. Oxford: Pergamon Press, 1964.

72 H. J. S. Smith, *Report on the Theory of Numbers*, 1894; reprinted by Chelsea Publishing Company, Inc., New York, 1965.

73 H. Stark, *An Introduction to Number Theory*. Chicago: Markham Publishing Company, 1970.

74 T. Storer, *Cyclotomy and Difference Sets*. Chicago: Markham Publishing Company, 1967.

75 R. Swan, Factorization of Polynomials over Finite Fields, *Pacific J. Math.*, **12**, 1962, 1099–1106.

76 E. Vegh, Primitive Roots Modulo a Prime as Consecutive Terms of an Arithmetic Progression, *J. Reine und Angew. Math.*, **235**, 1969, 185–188.

77 I. M. Vinogradov, *Elements of Number Theory*, trans. S. Kravetz. New York: Dover Publications, Inc., 1954.

78 E. Warning, Bemerkung zur vorstehenden Arbeit von Herrn Chevalley, *Abh. Math. Sem. Hamburg*, **11**, 1936, 76–83.

79 W. Waterhouse, The Sign of the Gauss Sum, *J. Number Theory*, **2**, No. 3, 1970, 363.

80 A. Weil, Number of Solutions of Equations in a Finite Field, *Bull. Am. Math. Soc.*, 1949, 497–508.

81 A. Weil, Jacobi Sums as "Grossencharaktere," *Trans. Am. Math. Soc.*, **73**, 1952, 487–495.

82 K. Yamamoto, On a Conjecture of Hasse Concerning Multiplicative Relations of Gauss Sums, *J. Combin. Theory*, **1**, 1966, 476–489.

83 A. Yokoyama, On the Gaussian Sum and the Jacobi Sum with Its Applications, *Tohoku Maths. J.* (2), **16**, 1964, 142–153.

SELECTED HINTS FOR THE EXERCISES

Chapter 1

6 Use Exercise 4.

8 Do it for the case $d = 1$ and then use Exercise 7 to do it in general.

9 Use Exercise 4.

15 Here is a generalization; a is an nth power iff $n | \text{ord}_p\, a$ for all primes p.

16 Use Exercise 15.

17 Use Exercise 15 to show that $a^2 = 2b^2$ implies that 2 is the square of an integer.

23 Begin by writing $4(a/2)^2 = (c - b)(c + b)$.

28 Show that $n^5 - n$ is divisible by 2, 3, and 5. Then use Exercise 9.

30 Let s be the largest integer such that $2^s \leq n$, and consider $\sum_{k=1}^{n} 2^{s-1}/k$. Show that this sum can be written in the form $a/b + \frac{1}{2}$ with b odd. Then use Exercise 29.

31 $2 = (1 + i)(1 - i) = -i(1 + i)^2$.

34 Since $\omega^2 = -1 - \omega$ we have $(1 - \omega)^2 = 1 - 2\omega + \omega^2 = -3\omega$, so $3 = -\omega^2(1 - \omega)^2$.

Chapter 2

1 Imitate the classical proof of Euclid.

2 Use $\text{ord}_p\,(a + b) \geq \min(\text{ord}_p\, a, \text{ord}_p\, b)$.

3 If $p_1, p_2, \ldots, p_t$ were all the primes, then $\varphi(p_1 p_2 \cdots p_t) = 1$. Now use the formula for φ and derive a contradiction.

5 Consider $2^2 + 1, 2^4 + 1, 2^8 + 1, \ldots$. No prime that divides one of these numbers can divide any other, by the previous exercise.

6 Count! Consider the set of pairs (s, t) with $p^s t \leq n$.

12 In each case the summand is multiplicative. Hence evaluate first at prime powers and then use multiplicativity.

17 Use the formula for $\sigma(n)$.

20 If $d | n$, then n/d also divides n.

22 If $(t, n) = 1$, then $(n - t, n) = 1$, so you can pair those numbers relatively prime to n in such a way that the sum of each pair is n.

Chapter 3

1 Suppose that $p_1, p_2, \ldots, p_t$ are all congruent to -1 modulo 6. Consider $N = 6p_1 p_2 \cdots p_t - 1$.

3 10^k is congruent to 1 modulo 3 and 9 and congruent to $(-1)^k$ modulo 11.

5 If a solution exists, then $x^3 \equiv 2\,(7)$ has a solution. Show that it doesn't.

10 If n is not a prime power, write $n = ab$ with $(a, b) = 1$. If $n = p^s$ with $s > 1$, then $(n - 1)!$ is divisible by $p \cdot p^{s-1} = p^s = n$. If $n = p^2$ and $p \neq 2$, then $(n - 1)!$ is divisible by $p \cdot 2p = 2n$.

13 Show that $n^p \equiv n\,(p)$ for all n by induction. If $(n, p) = 1$, then one can cancel n and get Fermat's formula.

17 Let x_i be a solution to $f(x) \equiv 0\,(p_i^{a_i})$ and solve the system $x \equiv x_i\,(p_i^{a_i})$.

23 Since $i \equiv -1\,(1 + i)$, we have $a + bi \equiv a - b\,(1 + i)$. Write $a - b = 2c + d$, where $d = 0$ or 1. Then $a + ib \equiv d\,(1 + i)$.

25 Write $\alpha = 1 + \beta\lambda$, cube both sides and take congruence modulo λ^4 to get $\alpha^3 \equiv 1 + (\beta^3 - \omega^2\beta)\lambda^3\,(\lambda^4)$. Then show that the term in parentheses is divisible by λ.

Chapter 4

4 If $(-a)^n \equiv 1$, and n is even, then $p - 1|n$. If n is odd, then $p - 1|2n$, which implies that $2|n$ is a contradiction.

6 This is a bit tricky. If 3 is not a primitive element, show that 3 is congruent to a square. Use Exercise 4 to show there is an integer a such that $-3 \equiv a^2\,(p)$. Now solve $2u \equiv -1 + a\,(p)$ and show that u has order 3. This would imply that $p = 1\,(3)$, which cannot be true.

7 Use the fact that 2 is not a square modulo p.

9 See Exercise 22 of Chapter 2 and use the fact that $g^{(p-1)/2} \equiv -1\,(p)$ for a primitive root g.

11 Express the numbers between 1 and $p - 1$ as the powers of a primitive root and use the formula for the sum of a geometric progression.

14 If $(ab)^s = e$, then $a^{ns} = 1$, implying that $m|ns$. Thus $m|s$. Similarly, $n|s$. Thus $mn|s$.

18 Choose a primitive element (e.g., 2) and construct the elements of order 7.

22 Show first that $1 + a + a^2 \equiv 0\,(p)$.

23 Use Proposition 4.2.1.

Chapter 5

3 Use the identity $4(ax^2 + bx + c) = (2ax + b)^2 - (b^2 - 4ac)$.

9 Using $k \equiv -(p - k)\,(p)$, show first that $2 \cdot 4 \cdot \ldots \cdot (p - 1) \equiv (-1)^{(p-1)/2} 1 \cdot 3 \cdot 5 \cdot \ldots \cdot p - 2\,(p)$.

10 Use Exercise 9.

13 If $x^4 - x^2 + 1 \equiv 0\,(p)$, then $(2x^2 - 1)^2 \equiv -3\,(p)$ and $(x^2 - 1)^2 \equiv -x^2\,(p)$. Conclude that $p = 1\,(3)$ and $p = 1\,(4)$ by using quadratic reciprocity.

18 Let $D = p_1 p_2 \cdots p_m$ and suppose that n is a nonresidue modulo p_i. Find a number b such that $b \equiv 1\,(p_i)$ and $b \equiv n\,(p_1)$ for $1 < i \leq m$. Then use the definition of the Jacobi symbol to show that $(b/D) = -1$.

23 Since $s^2 + 1 = (s + i)(s - i)$, if p is prime in $Z[i]$, then either $p|s + i$ or $p|s - i$, but neither alternative is true.

26 To prove (b) notice that $a + b$ is odd, so from $2p = (a + b)^2 + (a - b)^2$ we see that $(2p/a + b) = 1$. Now use the properties of the Jacobi symbol.

29 It is useful to consider the cases $p \equiv 1\,(4)$ and $p \equiv 3\,(4)$ separately.

30 To evaluate the sum notice that $(n(n + 1)/p) = ((2n + 1)^2 - 1/p)$.

Chapter 6

1 Find an equation of degree 4.

2 If $a_0\alpha^s + a_1\alpha^{s-1} + \cdots + a_s = 0$, with $a_i \in Z$, multiply both sides with a_0^{s-1} and conclude that $a_0\alpha$ is an algebraic integer.

3 Suppose that α and β satisfy monic equations with integer coefficients of degree m and n, respectively. Let γ be a root of $x^2 + \alpha x + \beta$ and show that the Z module generated by $\alpha^i\beta^j\gamma^k$, where $0 \leq i < m$, $0 \leq j < n$, and $k = 0$ or 1, is mapped into itself by γ.

10 Use $g_a = (a/p)g$ and the fact that $\sum_a (a/p) = 0$.

11 Remember that $1 + (t/p)$ is the number of solutions to $x^2 \equiv t\,(p)$ and that $\sum_t \zeta^t = 0$.

13 Use Exercise 12.

Chapter 7

3 Since $q \equiv 1\,(n)$, there are n solutions to $x^n = 1$. If $\beta^n = \alpha$, then the other solutions to $x^n = \alpha$ are given by $\gamma\beta$, where γ runs through the solutions of $x^n = 1$.

5 $q^n - 1 = (q - 1)(q^{n-1} + \cdots + q + 1)$. Since $q \equiv 1\,(n)$, we have $q^{n-1} + \cdots + q + 1 \equiv n \equiv 0\,(n)$. Thus $n(q - 1)$ divides $q^n - 1$.

7 Let $m = [K:F]$. α is a square in K iff $\alpha^{(q^m-1)/2} = 1$. If α is not a square in F, then $\alpha^{(q-1)/2} = -1$. Show that $\alpha^{(q^m-1)/2} = (-1)^m$. This formula yields the result.

9 Use the method of Exercise 7.

14 One can prove this by exactly the same method as for F_p. Alternatively, suppose that $q = p^m$. Let $f(x) \in F_p[x]$ be an irreducible of degree mn and let $g(x)$ be an irreducible factor of $f(x)$ in $F_q[x]$.

Let α be a root of $g(x)$ and show that $F_q \subset F_p(\alpha)$. Conclude that $F_q(\alpha) = F_p(\alpha)$ and that $[F_q(\alpha):F_q] = n$. It follows that $g(x)$ has degree n.

15 If $x^n - 1$ splits into linear factors in E, where $[E:F] = f$, then E has q^f elements and $n|q^f - 1$ since the roots of $x^n - 1$ form a subgroup of E^* of order n.

23 If β is a root of $x^p - x - \alpha$, then so are $\beta + 1$, $\beta + 2, \ldots,$ $\beta + (p - 1)$. Using this, one can show the statement about irreducibility. To prove the final assertion, notice that $\beta^p = \beta + \alpha$ implies that $\beta^{p^2} = \beta^p + \alpha^p = \beta + \alpha + \alpha^p$, etc. Thus $\beta^{p^n} = \beta +$ $\text{tr}(\alpha)$ and so $\beta \in F$ iff $\text{tr}(\alpha) = 0$.

Chapter 8

1 Use the Corollary to Proposition 8.1.3 and Proposition 8.1.4.

4 Make the substitution $t = (k/2)(u + 1)$ and use Exercise 3.

6 It follows from Exercise 5 together with part (d) of Theorem 1, or directly from Exercise 4 by substituting $k = 1$.

8 Use Proposition 8.1.5 and imitate the proof of Exercise 3.

14 Use Proposition 8.3.3.

19 First show that the number of solutions is given by $p^{r-1} +$ $J_0(\chi, \chi, \ldots, \chi)$, where χ is a character of order 2 and there are r components in J_0. Then use Proposition 8.5.1 and Theorem 3. Notice in particular that if r is odd, the answer is simply p^{r-1}.

Chapter 9

3 Use the fact that $N\gamma = a^2 - ab + b^2 \equiv 3(m + n) + 1\ (9)$.

4 Rewrite γ as $3(m + n) - 1 - 3n\lambda$. Thus $\gamma \equiv 3(m + n) - 1\ (3\lambda)$.

5 Remember that $3 = -\omega^2\lambda^2$.

7 $2 + 3\omega$, $-7 - 3\omega$, and $-4 - 3\omega$.

10 $D/5D$ has 25 elements. Thus $x^{24} - 1$ factors completely into linear factors in D.

13 Use Exercise 12 to show that the elements listed represent all the cubes in $D/5D$.

15 Remember that every element in $D/\pi D$ is represented by a rational integer.

19 Use Exercise 18, the law of cubic reciprocity, and induction on the number of primary primes dividing γ.

23 Let $p = \pi\bar{\pi}$, where π is primary. By Exercise 15 $x^3 \equiv 3\ (p)$ is solvable iff $\chi_\pi(3) = 1$. By Exercise 5 $\chi_\pi(3) = \omega^{2n}$, where $\pi =$ $a + b\omega$ and $b = 3n$. It follows that $x^3 \equiv 3\ (p)$ is solvable iff $9|b$.

Chapter 10

2 Map $\overline{(x_0, x_1, \ldots, x_{n-1})}$ to $\overline{(0, x_0, x_1, \ldots, x_{n-1})}$.

3 Since the number of points in $A^n(F)$ is q^n, the decomposition of $P^n(F)$ shows that the number of points in $P^n(F)$ is q^n plus the number of points in $P^{n-1}(F)$. One now proceeds by induction.

4 It is no loss of generality to assume that $a_0 \neq 0$. If $\overline{(x_0, x_1, \ldots, x_n)}$ is a solution, map it to the point $\overline{(x_1, x_2, \ldots, x_n)}$ of $P^{n-1}(F)$. Show this map is well defined, one to one, and onto.

5 Substitute, "dehomogenize," and use the fact that a polynomial of degree n has at most n roots.

9 The kth partial derivative is $ma_k x_k^{m-1}$. Since each $a_k \neq 0$ and m is prime to the characteristic, the only common zero of all the partial derivatives has all its components zero. This, however, does not correspond to a point of projective space.

12 The "homogenized" equation is $t^2x^2 + t^2y^2 + x^2y^2 = 0$. Setting $t = 0$ we see that the points at infinity are $(0, 0, 1)$ and $(0, 1, 0)$. Calculating partial derivatives and substituting shows that both these points are singular.

14 Consider the associated homogeneous equation and calculate the three partial derivatives. Assuming that a common solution exists, show that $4a^3 + 27b^2 = 0$.

19 The trace is identically zero on F_p iff $p|n$.

20 Consider the mapping $h(x) = x^p - x$ from F_q to F_q. Prove that it is a homomorphism and that its image has q/p elements. Prove also that the image of h is contained in the kernel of the trace mapping. Show that the latter map has less than or equal to q/p elements in its kernel. The result follows.

21 Count the number of such maps.

23 Substitute and calculate.

Chapter 11

4 In F_q there are $(q - 1)(2q + 1)$ points at infinity and q^2 finite points. Thus $N_s = 3p^{2s} - p^s - 1$.

7 The number of lines in $P^n(F)$ is equal to the number of planes $A^{n+1}(F)$ which pass through the origin. The answer is $(q^{n+1} - 1)(q^{n+1} - q)(q^2 - 1)^{-1}(q^2 - q)^{-1}$.

9 There is one point at infinity. For $x = 0$ there is only one point $(0, 0)$ on the curve. If $x \neq 0$, let $t = y/x$ and consider $t^2 = x + 1$. This has $p - 2$ solutions with $x \neq 0$. Altogether there are p solutions in F_p. Similarly, there are q solutions in F_q. Thus the answer is $(1 - pu)^{-1}$.

12 To begin with, calculate the number of solutions to $u^2 - v^4 = 4D$.

16 The important facts are that $N_{F_s/F}$ is a homomorphism which is onto, and that the group of multiplicative characters of a finite field is cyclic.

18 Use the relation between Gauss sums and Jacobi sums and the Hasse–Davenport relation.

19 After expanding the terms of the product into geometric series, the result reduces to the fact that every monic polynomial is the product of monic irreducible polynomials in a unique way.

20 Use the identity $1 - T^s = \prod_{k=0}^{s-1} (1 - \zeta^k T)$, where $\zeta = e^{2\pi i/s}$.

INDEX